DEATH
BY
BLACK
H ● LE

Origins: Fourteen Billion Years of Cosmic Evolution
(with Donald Goldsmith)

My Favorite Universe
(A 12-Part Video Lecture Series)

One Universe: At Home in the Cosmos
(with Charles Liu & Robert Irion)

Cosmic Horizons: Astronomy at the Cutting Edge
(with Steven Soter, eds)

The Sky Is Not the Limit:
Adventures of an Urban Astrophysicist

Just Visiting This Planet

Universe Down to Earth

Merlin's Tour of the Universe

NEIL DeGRASSE TYSON

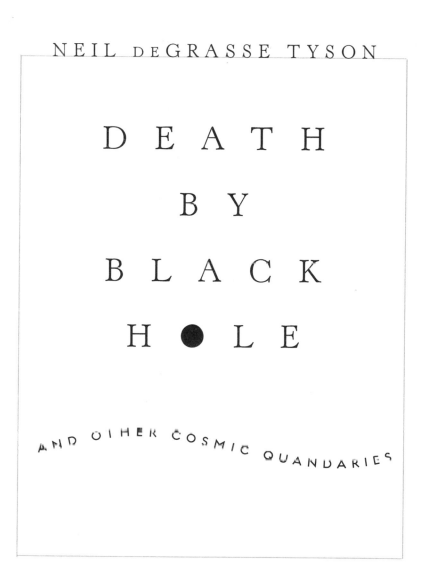

DEATH

BY

BLACK

H●LE

AND OTHER COSMIC QUANDARIES

W. W. NORTON & COMPANY

New York · *London*

For information about permission to reproduce selections from this book, write to
Permissions, W. W. Norton & Company, Inc., 500 Fifth Avenue, New York, NY 10110

Manufacturing by R.R. Donnelley, Bloomsburg Division
Book design by JAM Design
Production manager: Amanda Morrison

Library of Congress Cataloging-in-Publication Data
Tyson, Neil deGrasse.
Death by black hole : and other cosmic quandaries / Neil deGrasse Tyson. — 1st ed.
p. cm.
Includes bibliographical references and indexes.
ISBN-13: 978-0-393-06224-3
ISBN-10: 0-393-06224-4
1. Black holes (Astronomy) 2. Cosmology. 3. Astrobiology. 4. Solar system.
5. Religion and science. I. Title.
QB843.B55T97 2007
523.8'875—dc22

2006022058

W. W. Norton & Company, Inc., 500 Fifth Avenue, New York, N.Y. 10110
www.wwnorton.com

W. W. Norton & Company Ltd., Castle House, 75/76 Wells Street, London W1T 3QT

4 5 6 7 8 9 0

CONTENTS

SECTION 3 WAYS AND MEANS OF NATURE
How Nature presents herself to the inquiring mind

SECTION 4 THE MEANING OF LIFE
The challenges and triumphs of knowing how we got here

SECTION 5 WHEN THE UNIVERSE TURNS BAD
All the ways the cosmos wants to kill us

My own suspicion is that the Universe

is not only queerer than we suppose,

but queerer than we can suppose.

—J. B. S. Haldane
Possible Worlds (1927)

.

PREFACE

I see the universe not as a collection of objects, theories, and phenomena, but as a vast stage of actors driven by intricate twists of story line and plot. So when writing about the cosmos, it feels natural to bring readers into the theater, behind the scenes, to see up close for themselves what the set designs look like, how the scripts were written, and where the stories will go next. My goal at all times is to communicate insight into how the universe works, which is harder than the simple conveyance of facts. Times arise along the way, as for the drama icon itself, to smile or to frown when the cosmos calls for it. Times arise to be scared witless when the cosmos calls for that, too. So I think of *Death by Black Hole* as a reader's portal to all that moves, enlightens, and terrifies us in the universe.

Each chapter first appeared, in one form or another, on the pages of *Natural History* magazine under the heading "Universe" and span the 11-year period of 1995 through 2005. *Death by Black Hole* forms a kind of "Best of the Universe" and includes some of the most requested essays I have written, mildly edited for continuity and to reflect emergent trends in science.

I submit this collection to you, the reader, for what might be a welcome diversion from your day's routine.

Neil deGrasse Tyson
New York City
October 2006

ACKNOWLEDGMENTS

My formal expertise in the universe concerns stars, stellar evolution, and galactic structure. And so I could not possibly write with authority about the breadth of subjects in this collection without the careful eyes of colleagues whose comments on my monthly manuscripts often made the difference between a simple idea described and an idea nuanced with meaning drawn from the frontier of cosmic discovery. For matters regarding the solar system, I am grateful to Rick Binzel, my former classmate in graduate school and now professor of Planetary Sciences at MIT. He has received many a phone call from me, in desperate search of a reality-check on what I had written or what I had planned to write about the planets and their environments.

Others in this role include Princeton Astrophysics Professors Bruce Draine, Michael Strauss, and David Spergel whose collective expertise in cosmo-chemistry, galaxies, and cosmology allowed me to reach deeper into that store of cosmic places than would otherwise be possible. Among my colleagues, the ones who are closest to these essays include Princeton's Robert Lupton, who, being properly educated in England, looks to me as though he knows everything about everything. For most of the essays in this volume, Robert's remarkable attention to scientific as well as literary detail provided reliable monthly enhancement to whatever I had penned. Another colleague and generalist who keeps watch over my work is

Steven Soter. My writings are somehow incomplete without first passing them to his attention.

From the literary world, Ellen Goldensohn, who was my first editor at *Natural History* magazine, invited me to write a column in 1995 after hearing me interviewed on National Public Radio. I agreed on the spot. And this monthly task remains one of the most exhausting and exhilarating things I do. Avis Lang, my current editor continues the effort begun by Ellen, ensuring that, without compromise, I say what I mean and mean what I say. I am indebted to both of them for the time they have invested to make me be a better writer. Others who have helped to improve or otherwise enhance the content of one or more essays include Phillip Branford, Bobby Fogel, Ed Jenkins, Ann Rae Jonas, Betsy Lerner, Mordecai Mark Mac-Low, Steve Napear, Michael Richmond, Bruce Stutz, Frank Summers, and Ryan Wyatt. Hayden volunteer Kyrie Bohin-Tinch made a heroic first pass at helping me to organize the universe of this book. And I offer further thanks to Peter Brown, editor-in-chief of *Natural History* magazine, for his overall support of my writing efforts and for granting permission to reproduce the essays of my choice for this collection.

This page would be incomplete without a brief expression of debt to Stephen Jay Gould, whose *Natural History* column "This View of Life" ran for three hundred essays. We overlapped at the magazine for seven years, from 1995 through 2001, and not a month passed where I did not feel his presence. Stephen practically invented the modern essay form, and his influence on my work is manifest. Wherever I am compelled to reach deep into the history of science, I would acquire and turn the fragile pages of rare books from centuries past, as Gould so often did, drawing from them a rich sampling of how those who came before us attempted to understand the operations of the natural world. His premature death at age 60, like that of Carl Sagan at age 62, left a vacuum in the world of science communication that remains to this day unfilled.

PROLOGUE:
The Beginning of Science

The success of known physical laws to explain the world around us has consistently bred some confident and cocky attitudes toward the state of human knowledge, especially when the holes in our knowledge of objects and phenomena are perceived to be small and insignificant. Nobel laureates and other esteemed scientists are not immune from this stance, and in some cases have embarrassed themselves.

A famous end-of-science prediction came in 1894, during the speech given by the soon-to-be Nobel laureate Albert A. Michelson on the dedication of the Ryerson Physics Lab, at the University of Chicago:

> *The more important fundamental laws and facts of physical science have all been discovered, and these are now so firmly established that the possibility of their ever being supplanted in consequence of new discoveries is exceedingly remote. . . . Future discoveries must be looked for in the sixth place of decimals.* (Barrow 1988, p. 173)

One of the most brilliant astronomers of the time, Simon Newcomb, who was also cofounder of the American Astronomical Society, shared Michelson's views in 1888 when he noted, "We are probably nearing the limit of all we can know about astronomy" (1888, p. 65). Even the great physicist Lord Kelvin, who, as we shall see in Section 3, had the absolute temperature scale named after him, fell victim to his own confidence in 1901 with the claim, "There is nothing new to be discovered in physics now. All that remains is more and more precise measurement" (1901, p. 1). These comments were expressed at a time when the luminiferous ether was still the presumed medium in which light propagated through space, and when the slight difference between the observed and predicted path of Mercury around the Sun was real and unsolved. These quandaries were perceived at the time to be small, requiring perhaps only mild adjustments to the known physical laws to account for them.

Fortunately, Max Planck, one of the founders of quantum mechanics, had more foresight than his mentor. Here, in a 1924 lecture, he reflects on the advice given to him in 1874:

> *When I began my physical studies and sought advice from my venerable teacher Philipp von Jolly . . . he portrayed to me physics as a highly developed, almost fully matured science. . . . Possibly in one or another nook there would perhaps be a dust particle or a small bubble to be examined and classified, but the system as a whole stood there fairly secured, and theoretical physics approached visibly that degree of perfection which, for example, geometry has had already for centuries.* (1996, p. 10)

Initially Planck had no reason to doubt his teacher's views. But when our classical understanding of how matter radiates energy could not be reconciled with experiment, Planck became a reluctant revolutionary in 1900 by suggesting the existence of the quantum, an indivisible unit of energy that heralded an era of new physics. The next 30 years would see the discovery of the special and general theories of relativity, quantum mechanics, and the expanding universe.

With all this myopic precedence you would think that the brilliant and prolific physicist Richard Feynman would have known better. In his charming 1965 book *The Character of Physical Law*, he declares:

We are very lucky to be living in an age in which we are still making discoveries. . . . The age in which we live is the age in which we are discovering the fundamental laws of nature, and that day will never come again. It is very exciting, it is marvelous, but this excitement will have to go. (Feynman 1994, p. 166)

I claim no special knowledge of when the end of science will come, or where the end might be found, or whether an end exists at all. What I do know is that our species is dumber than we normally admit to ourselves. This limit of our mental faculties, and not necessarily of science itself, ensures to me that we have only just begun to figure out the universe.

Let's assume, for the moment, that human beings are the smartest species on Earth. If, for the sake of discussion, we define "smart" as the capacity of a species to do abstract mathematics then one might further assume that human beings are the only smart species to have ever lived.

What are the chances that this first and only smart species in the history of life on Earth has enough smarts to completely figure out how the universe works? Chimpanzees are an evolutionary hair's-width from us yet we can agree that no amount of tutelage will ever leave a chimp fluent in trigonometry. Now imagine a species on Earth, or anywhere else, as smart compared with humans as humans are compared with chimpanzees. How much of the universe might they figure out?

Tic-tac-toe fans know that the game's rules are sufficiently simple that it's possible to win or tie every game—if you know which first-moves to make. But young children play the game as though the outcome were remote and unknowable. The rules of engagement are also clear and simple for the game of chess, but the challenge of predicting your opponent's upcoming sequence of moves grows exponentially as the game proceeds. So adults—even smart and talented ones—are challenged by the game and play it as though the end were a mystery.

Let's go to Isaac Newton, who leads my list of the smartest people who ever lived. (I am not alone here. A memorial inscription on a bust of him in Trinity College, England, proclaims *Qui genus humanum ingenio superavit,* which loosely translates from the Latin to "of all humans, there is no greater intellect.") What did Newton observe about his state of knowledge?

> *I do not know what I appear to the world; but to myself I seem to have been only like a boy playing on a seashore, and diverting myself in now and then finding a smoother pebble or a prettier shell than ordinary, whilst the great ocean of truth lay undiscovered before me.*
> (Brewster 1860, p. 331)

The chessboard that is our universe has revealed some of its rules, but much of the cosmos still behaves mysteriously—as though there remain secret, hidden regulations to which it abides. These would be rules not found in the rule book we have thus far written.

The distinction between knowledge of objects and phenomena, which operate within the parameters of known physical laws, and knowledge of the physical laws themselves is central to any perception that science might be coming to an end. The discovery of life on the planet Mars, or beneath the floating ice sheets of Jupiter's moon Europa, would be the greatest discovery of any kind ever. You can bet, however, that the physics and chemistry of its atoms will be the same as the physics and chemistry of atoms here on Earth. No new laws necessary.

But let's peek at a few unsolved problems from the underbelly of modern astrophysics that expose the breadth and depth of our contemporary ignorance, the solutions of which, for all we know, await the discovery of entirely new branches of physics.

While our confidence in the big bang description of the origin of the universe is very high, we can only speculate what lies beyond our cosmic horizon, 13.7 billion light-years from us. We can only guess what happened before the big bang or why there should have been a big bang in the first place. Some predictions, from the limits of quantum mechanics, allow our expanding universe to be the result of just one fluctuation from a primordial space-time foam, with countless other fluctuations spawning countless other universes.

Shortly after the big bang, when we try to get our computers to make the universe's hundred billion galaxies, we have trouble simultaneously matching the observational data from early and late times in the universe. A coherent description of the formation and evolution of the large-scale structure of the universe continues to elude us. We seem to be missing some important pieces of the puzzle.

Newton's laws of motion and gravity looked good for hundreds of years, until they needed to be modified by Einstein's theories of motion and gravity—the relativity theories. Relativity now reigns supreme. Quantum mechanics, the description of our atomic and nuclear universe, also reigns supreme. Except that as conceived, Einstein's theory of gravity is irreconcilable with quantum mechanics. They each predict different phenomena for the domain in which they might overlap. Something's got to surrender. Either there's a missing part of Einstein's gravity that enables it to accept the tenets of quantum mechanics, or there's a missing part of quantum mechanics that enables it to accept Einstein's gravity.

Perhaps there's a third option: the need for a larger, inclusive theory that supplants them both. Indeed, string theory has been invented and called upon to do just that. It attempts to reduce the existence of all matter, energy, and their interactions to the simple existence of higher dimensional vibrating strings of energy. Different modes of vibration would reveal themselves in our measly dimensions of space and time as different particles and forces. Although string theory has had its adherents for more than 20 years, its claims continue to lie outside our current experimental capacity to verify its formalisms. Skepticism is rampant, but many are nonetheless hopeful.

We still do not know what circumstances or forces enabled inanimate matter to assemble into life as we know it. Is there some mechanism or law of chemical self-organization that escapes our awareness because we have nothing with which to compare our

Earth-based biology, and so we cannot evaluate what is essential and what is irrelevant to the formation of life?

We've known since Edwin Hubble's seminal work during the 1920s that the universe is expanding, but we've only just learned that the universe is also accelerating, by some antigravity pressure dubbed "dark energy" for which we have no working hypothesis to understand.

At the end of the day, no matter how confident we are in our observations, our experiments, our data, or our theories, we must go home knowing that 85 percent of all the gravity in the cosmos comes from an unknown, mysterious source that remains completely undetected by all means we have ever devised to observe the universe. As far as we can tell, it's not made of ordinary stuff such as electrons, protons, and neutrons, or any form of matter or energy that interacts with them. We call this ghostly, offending substance "dark matter," and it remains among the greatest of all quandaries.

Does any of this sound like the end of science? Does any of this sound like we are on top of the situation? Does any of this sound like it's time to congratulate ourselves? To me it sounds like we are all helpless idiots, not unlike our kissing cousin, the chimpanzee, trying to learn the Pythagorean theorem.

Maybe I'm being a little hard on *Homo sapiens* and have carried the chimpanzee analogy a little too far. Perhaps the question is not how smart is an individual of a species, but how smart is the collective brain-power of the entire species. Through conferences, books, other media, and of course the Internet, humans routinely share their discoveries with others. While natural selection drives Darwinian evolution, the growth of human culture is largely Lamarckian, where new generations of humans inherit the acquired discoveries of generations past, allowing cosmic insight to accumulate without limit.

Each discovery of science therefore adds a rung to a ladder of knowledge whose end is not in sight because we are building the ladder as we go along. As far as I can tell, as we assemble and ascend this ladder, we will forever uncover the secrets of the universe—one by one.

DEATH
BY
BLACK
H●LE

SECTION 1

THE NATURE OF KNOWLEDGE

THE CHALLENGES OF KNOWING
WHAT IS KNOWABLE IN THE UNIVERSE

COMING TO OUR SENSES

Equipped with his five senses, man explores the universe around him and calls the adventure science.
—EDWIN P. HUBBLE (1889–1953), *The Nature of Science*

Among our five senses, sight is the most special to us. Our eyes allow us to register information not only from across the room but also from across the universe. Without vision, the science of astronomy would never have been born and our capacity to measure our place in the universe would have been hopelessly stunted. Think of bats. Whatever bat secrets get passed from one generation to the next, you can bet that none of them is based on the appearance of the night sky.

When thought of as an ensemble of experimental tools, our senses enjoy an astonishing acuity and range of sensitivity. Our ears can register the thunderous launch of the space shuttle, yet they can also hear a mosquito buzzing a foot away from our head. Our sense of touch allows us to feel the magnitude of a bowling ball dropped on our big toe, just as we can tell when a one-milligram bug crawls along our arm. Some people enjoy munching on habanero peppers while sensitive tongues can identify the presence of food flavors on the level of parts per million. And our eyes can register the bright sandy terrain on a sunny beach, yet these same eyes have no trouble spotting a lone match, freshly lit, hundreds of feet across a darkened auditorium.

But before we get carried away in praise of ourselves, note that what we gain in breadth we lose in precision: we register the world's stimuli in logarithmic rather than linear increments. For example, if you increase the energy of a sound's volume by a factor of 10, your ears will judge this change to be rather small. Increase it by a factor of 2 and you will barely take notice. The same holds for our capacity to measure light. If you have ever viewed a total solar eclipse you may have noticed that the Sun's disk must be at least 90 percent covered by the Moon before anybody comments that the sky has darkened. The stellar magnitude scale of brightness, the well-known acoustic decibel scale, and the seismic scale for earthquake severity are each logarithmic, in part because of our biological propensity to see, hear, and feel the world that way.

WHAT, IF ANYTHING, lies beyond our senses? Does there exist a way of knowing that transcends our biological interfaces with the environment?

Consider that the human machine, while good at decoding the basics of our immediate environment—like when it's day or night or when a creature is about to eat us—has very little talent for decoding how the rest of nature works without the tools of science. If we want to know what's out there then we require detectors other than the ones we are born with. In nearly every case, the job of a scientific apparatus is to transcend the breadth and depth of our senses.

Some people boast of having a sixth sense, where they profess to know or see things that others cannot. Fortune-tellers, mind readers, and mystics are at the top of the list of those who lay claim to mysterious powers. In so doing, they instill widespread fascination in others, especially book publishers and television producers. The questionable field of parapsychology is founded on the expectation that at least some people actually harbor such talents. To me, the biggest mystery of them all is why so many fortune-telling psychics choose to work the phones on TV hotlines instead of becoming insanely wealthy trading futures contracts on Wall Street. And here's a news headline none of us has seen, "Psychic Wins the Lottery."

Quite independent of this mystery, the persistent failures of controlled, double-blind experiments to support the claims of parapsychology suggest that what's going on is nonsense rather than sixth sense.

On the other hand, modern science wields dozens of senses. And scientists do not claim these to be the expression of special powers, just special hardware. In the end, of course, the hardware converts the information gleaned from these extra senses into simple tables, charts, diagrams, or images that our inborn senses can interpret. In the original *Star Trek* sci-fi series, the crew that beamed down from their starship to the uncharted planet always brought with them a tricorder—a handheld device that could analyze anything they encountered, living or inanimate, for its basic properties. As the tricorder was waved over the object in question, it made an audible spacey sound that was interpreted by the user.

Suppose a glowing blob of some unknown substance were parked right in front of us. Without some diagnostic tool like a tricorder to help, we would be clueless to the blob's chemical or nuclear composition. Nor could we know whether it has an electromagnetic field, or whether it emits strongly in gamma rays, x-rays, ultraviolet, microwaves, or radio waves. Nor could we determine the blob's cellular or crystalline structure. If the blob were far out in space, appearing as an unresolved point of light in the sky, our five senses would offer us no insight to its distance, velocity though space, or its rate of rotation. We further would have no capacity to see the spectrum of colors that compose its emitted light, nor could we know whether the light is polarized.

Without hardware to help our analysis, and without a particular urge to lick the stuff, all we can report back to the starship is, "Captain, it's a blob." Apologies to Edwin P. Hubble, the quote that opens this chapter, while poignant and poetic, should have instead been:

Equipped with our five senses, along with telescopes and microscopes and mass spectrometers and seismographs and magnetometers and particle accelerators and detectors across the electromagnetic spectrum, we explore the universe around us and call the adventure science.

Think of how much richer the world would appear to us and how much earlier the nature of the universe would have been discovered if we were born with high-precision, tunable eyeballs. Dial up the radio-wave part of the spectrum and the daytime sky becomes as dark as night. Dotting that sky would be bright and famous sources of radio waves, such as the center of the Milky Way, located behind some of the principal stars of the constellation Sagittarius. Tune into microwaves and the entire cosmos glows with a remnant from the early universe, a wall of light set forth 380,000 years after the big bang. Tune into x-rays and you immediately spot the locations of black holes, with matter spiraling into them. Tune into gamma rays and see titanic explosions scattered throughout the universe at a rate of about one per day. Watch the effect of the explosion on the surrounding material as it heats up and glows in other bands of light.

If we were born with magnetic detectors, the compass would never have been invented because we wouldn't ever need one. Just tune into Earth's magnetic field lines and the direction of magnetic north looms like Oz beyond the horizon. If we had spectrum analyzers within our retinas, we would not have to wonder what we were breathing. We could just look at the register and know whether the air contained sufficient oxygen to sustain human life. And we would have learned thousands of years ago that the stars and nebulae in the Milky Way galaxy contain the same chemical elements found here on Earth.

And if we were born with big eyes and built-in Doppler motion detectors, we would have seen immediately, even as grunting troglodytes, that the entire universe is expanding—with distant galaxies all receding from us.

If our eyes had the resolution of high-performance microscopes, nobody would have ever blamed the plague and other sicknesses on divine wrath. The bacteria and viruses that made us sick would be in plain view as they crawled on our food or as they slid through open wounds in our skin. With simple experiments, we could easily tell which of these bugs were bad and which were good. And of course postoperative infection problems would have been identified and solved hundreds of years earlier.

If we could detect high-energy particles, we would spot radioactive substances from great distances. No Geiger counters necessary. We could even watch radon gas seep through the basement floor of our homes and not have to pay somebody to tell us about it.

THE HONING OF our senses from birth through childhood allows us, as adults, to pass judgment on events and phenomena in our lives, declaring whether they "make sense." Problem is, hardly any scientific discoveries of the past century flowed from the direct application of our five senses. They flowed instead from the direct application of sense-transcendent mathematics and hardware. This simple fact is entirely responsible for why, to the average person, relativity, particle physics, and 10-dimensional string theory make no sense. Include in the list black holes, wormholes, and the big bang. Actually, these ideas don't make much sense to scientists either, or at least not until we have explored the universe for a long time, with all the senses that are technologically available. What emerges, eventually, is a newer and higher level of "common sense" that enables a scientist to think creatively and to pass judgment in the unfamiliar underworld of the atom or in the mind-bending domain of higher-dimensional space. The twentieth-century German physicist Max Planck made a similar observation about the discovery of quantum mechanics:

Modern Physics impresses us particularly with the truth of the old doctrine which teaches that there are realities existing apart from our sense-perceptions, and that there are problems and conflicts where these realities are of greater value for us than the richest treasures of the world of experience. (1931, p. 107)

Our five senses even interfere with sensible answers to stupid metaphysical questions like, "If a tree falls in the forest and nobody is around to hear it, does it make a sound?" My best answer is, "How do you know it fell?" But that just gets people angry. So I offer a senseless analogy, "Q: If you can't smell the carbon monoxide, then how do you know it's there? A: You drop dead." In modern times, if the sole measure of what's out there flows from your five senses then a precarious life awaits you.

Discovering new ways of knowing has always heralded new windows on the universe that tap into our growing list of nonbiological senses. Whenever this happens, a new level of majesty and complexity in the universe reveals itself to us, as though we were technologically evolving into supersentient beings, always coming to our senses.

ON EARTH
AS IN THE HEAVENS

Until Isaac Newton wrote down the universal law of gravitation, there was little reason to presume that the laws of physics on Earth were the same as everywhere else in the universe. Earth had earthly things going on and the heavens had heavenly things going on. Indeed, according to many scholars of the day, the heavens were unknowable to our feeble, mortal minds. As further detailed in Section 7, when Newton breached this philosophical barrier by rendering all motion comprehensible and predictable, some theologians criticized him for leaving nothing for the Creator to do. Newton had figured out that the force of gravity pulling ripe apples from their branches also guides tossed objects along their curved trajectories and directs the Moon in its orbit around Earth. Newton's law of gravity also guides planets, asteroids, and comets in their orbits around the Sun and keeps hundreds of billions of stars in orbit within our Milky Way galaxy.

This universality of physical laws drives scientific discovery like nothing else. And gravity was just the beginning. Imagine the excitement among nineteenth-century astronomers when laboratory prisms, which break light beams into a spectrum of colors, were first turned to the Sun. Spectra are not only beautiful but also contain oodles of information about the light-emitting object, including its temperature and composition. Chemical elements reveal themselves by their unique patterns of light or dark bands that cut across the spectrum. To people's delight and amazement, the chemical signatures on the Sun were identical to those in the laboratory. No longer the exclusive tool of chemists, the prism showed that as different as the Sun is from Earth in size, mass, temperature, location, and appearance, both contained the same stuff—hydrogen, carbon, oxygen, nitrogen, calcium, iron, and so forth. But more important than a laundry list of shared ingredients was the recognition that whatever laws of physics prescribed the formation of these spectral signatures on the Sun, the same laws were operating on Earth, 93 million miles away.

So fertile was this concept of universality that it was successfully applied in reverse. Further analysis of the Sun's spectrum revealed the signature of an element that had no known counterpart on Earth. Being of the Sun, the new substance was given a name derived from the Greek word *helios* (the Sun). Only later was it discovered in the lab. Thus, "helium" became the first and only element in the chemist's periodic table to be discovered someplace other than Earth.

OKAY, THE LAWS of physics work in the solar system, but do they work across the galaxy? Across the universe? Across time itself? Step by step, the laws were tested. The nearby stars also revealed familiar chemicals. Distant binary stars, bound in mutual orbit, seem to know all about Newton's laws of gravity. For the same reason, so do binary galaxies.

And, like the geologist's stratified sediments, the farther away we look, the further back in time we see. Spectra from the most distant objects in the universe show the same chemical signatures that we see everywhere else in the universe. True, heavy elements were less abundant back then—they are manufactured primarily in subsequent generations of exploding stars—but the laws describing the atomic and molecular process that created these spectral signatures remain intact.

Of course, not all things and phenomena in the cosmos have

counterparts on Earth. You've probably never walked through a cloud of glowing million-degree plasma, and you've probably never stumbled upon a black hole on the street. What matters is the universality of the laws of physics that describe them. When spectral analysis was first turned to the light emitted by interstellar nebulae, an element appeared that, once again, had no counterpart on Earth. But the periodic table of elements had no missing boxes; when helium was discovered there were several. So astrophysicists invented the name "nebulium" as a placeholder, until they could figure out what was going on. Turned out that in space, gaseous nebulae are so rarefied that atoms go long stretches without colliding with each other. Under these conditions, electrons can do things within atoms that had never before been seen in Earth labs. Nebulium was simply the signature of ordinary oxygen doing extraordinary things.

This universality of physical laws tells us that if we land on another planet with a thriving alien civilization, they will be running on the same laws that we have discovered and tested here on Earth—even if the aliens harbor different social and political beliefs. Furthermore, if you wanted to talk to the aliens, you can bet they don't speak English or French or even Mandarin Chinese. You don't even know whether shaking their hands—if indeed they have hands to shake—would be considered and act of war or of peace. Your best hope is to find a way to communicate using the language of science.

Such an attempt was made in the 1970s with the *Pioneer 10* and *11* and *Voyager 1* and *2* spacecraft, the only ones given a great enough speed to escape the solar system's gravitational pull. *Pioneer* donned a golden etched plaque that showed, in pictograms, the layout of our solar system, our location in the Milky Way galaxy, and the structure of the hydrogen atom. *Voyager* went further and included diverse sounds from mother Earth including the human heartbeat, whale "songs," and musical selections ranging from the works of Beethoven to Chuck Berry. While this humanized the message, it's not clear whether alien ears would have a clue what they were listening to—assuming they have ears in the first place. My favorite parody of this gesture was a skit on *Saturday Night Live,* appearing shortly after the *Voyager* launch. NASA receives a reply from the aliens who recovered the spacecraft. The note simply requests, "Send more Chuck Berry."

AS WE WILL see in great detail in Section 3, science thrives not only on the universality of physical laws but also on the existence and persistence of physical constants. The constant of gravitation, known by most scientists as "big G," supplies Newton's equation of gravity with the measure of how strong the force will be, and has been implicitly tested for variation over eons. If you do the math, you can determine that a star's luminosity is steeply dependent on big G. In other words, if big G had been even slightly different in the past, then the energy output of the Sun would have been far more variable than anything that the biological, climatological, or geological records indicate. In fact, no time-dependent or location-dependent fundamental constants are known—they appear to be truly constant.

Such are the ways of our universe.

Among all constants, the speed of light is surely the most famous. No matter how fast you go, you will never overtake a beam of light. Why not? No experiment ever conducted has ever revealed an object of any form reaching the speed of light. Well-tested laws of physics predict and account for this. These statements sound closed-minded. True, some of the most embarrassing science-based proclamations in the past have underestimated the ingenuity of inventors and engineers: "We will never fly." "Flying will never be commercially feasible." "We will never fly faster than sound." "We will never split the atom." "We will never go to the Moon." You've heard them. What they have in common is that no established law of physics stood in their way.

The claim "We will never outrun a beam of light" is a qualitatively different prediction. It flows from basic, time-tested physical principles. No doubt about it. Highway signs for interstellar travelers of the future will surely read:

The Speed of Light:
It's Not Just a Good Idea
It's the Law.

The good thing about the laws of physics is that they require no law enforcement agencies to maintain them, although I once owned a nerdy T-shirt that loudly proclaimed, "OBEY GRAVITY."

Many natural phenomena reflect the interplay of multiple physical laws operating at once. This fact often complicates the analysis and, in most cases, requires supercomputers to calculate things and to keep track of important parameters. When comet Shoemaker-Levy 9 plunged into and then exploded within Jupiter's gas-rich atmosphere in 1994, the most accurate computer model of what was to happen combined the laws of fluid mechanics, thermodynamics, kinematics, and gravitation. Climate and weather represent other leading examples of complicated (and difficult-to-predict) phenomena. But the basic laws governing them are still at work. Jupiter's Great Red Spot, a raging anticyclone that has been going strong for at least 350 years, is driven by the identical physical processes that generate storms on Earth and elsewhere in the solar system.

THE CONSERVATION LAWS, where the amount of some measured quantity remains unchanged *no matter what* are another class of universal truths. The three most important are the conservation of mass and energy, the conservation of linear and angular momentum, and the conservation of electric charge. These laws are in evidence on Earth and everywhere we have thought to look in the universe—from the domain of particle physics to the large-scale structure of the universe.

In spite of all this boasting, all is not perfect in paradise. As already noted, we cannot see, touch, or taste the source of 85 percent of the gravity of the universe. This mysterious dark matter, which remains undetected except for its gravitational pull on matter we see, may be composed of exotic particles that we have yet to discover or identify. A tiny subset of astrophysicists, however, remain unconvinced and have suggested that dark matter does not exist—you simply need to modify Newton's law of gravity. Just add a few components to the equations and all will be well.

Perhaps one day we will learn that Newton's gravity indeed requires adjustment. That'll be okay. It has happened once before. In 1916, Albert Einstein published his general theory of relativity, which reformulated the principles of gravity in a way that applied to objects of extremely high mass, a realm unknown to Newton, and where his law of gravity breaks down. The lesson? Our confidence flows through the range of conditions over which a law has been tested and verified. The broader this range, the more powerful the law becomes in describing the cosmos. For ordinary household gravity, Newton's law works just fine. For black holes and the large-scale structure of the universe, we need general relativity. They each work flawlessly in their own domain, wherever that domain may be in the universe.

TO THE SCIENTIST, the universality of physical laws makes the cosmos a marvelously simple place. By comparison, human nature—the psychologist's domain—is infinitely more daunting. In America, school boards vote on the subjects to be taught in the classroom, and in some cases these votes are cast according to the whims of social and political tides or religious philosophies. Around the world, varying belief systems lead to political differences that are not always resolved peacefully. And some people talk to bus stop stanchions. The remarkable feature of physical laws is that they apply everywhere, whether or not you choose to believe in them. After the laws of physics, everything else is opinion.

Not that scientists don't argue. We do. A lot. When we do, however, we are usually expressing opinions about the interpretation of ratty data on the frontier of our knowledge. Wherever and whenever a physical law can be invoked in the discussion, the debate is guaranteed to be brief: No, your idea for a perpetual motion machine will never work—it violates laws of thermodynamics. No, you can't build a time machine that will enable you to go back and kill your mother before you were born—it violates causality laws. And without violating momentum laws, you cannot spontaneously levitate and hover above the ground, whether or not you are seated in the lotus position. Although, in principle, you could perform this stunt if you managed to let loose a powerful and sustained exhaust of flatulents.

Knowledge of physical laws can, in some cases, give you the confidence to confront surly people. A few years ago I was having a hot-cocoa nightcap at a dessert shop in Pasadena, California. I had ordered it with whipped cream, of course. When it arrived at the table, I saw no trace of the stuff. After I told the waiter that my cocoa was plain, he asserted I couldn't see the whipped cream because it sank to the bottom. Since whipped cream has a very low density and floats on all liquids that humans consume, I offered the waiter two possible explanations: either somebody forgot to add the whipped cream to my hot cocoa or the universal laws of physics were different in his restaurant. Unconvinced, he brought over a dollop of whipped cream to test for himself. After bobbing once or twice in my cup, the whipped cream sat up straight and afloat.

What better proof do you need of the universality of physical law?

THREE

SEEING ISN'T BELIEVING

S o much of the universe appears to be one way but is really another that I wonder, at times, whether there's an ongoing conspiracy designed to embarrass astrophysicists. Examples of such cosmic tomfoolery abound.

In modern times we take for granted that we live on a spherical planet. But the evidence for a flat Earth seemed clear enough for thousands of years of thinkers. Just look around. Without satellite imagery, it's hard to convince yourself that the Earth is anything but flat, even when you look out of an airplane window. What's true on Earth is true on all smooth surfaces in non-Euclidean geometry: a sufficiently small region of any curved surface is indistinguishable from a flat plane. Long ago, when people did not travel far from their birthplace, a flat Earth supported the ego-stroking view that your hometown occupied the exact center of Earth's surface and that all points along the horizon (the edge of your world) were equally distant from you. As one might expect, nearly every map of a flat Earth depicts the map-drawing civilization at its center.

Now look up. Without a telescope, you can't tell how far away the stars are. They keep their places, rising and setting as if they were glued to the inside surface of a dark, upside-down cereal bowl. So why not assume all stars to be the same distance from Earth, whatever that distance might be?

But they're not all equally far away. And of course there is no bowl. Let's grant that the stars are scattered through space, hither and yon. But how hither, and how yon? To the unaided eye the brightest stars are more than a hundred times brighter than the dimmest. So the dim ones are obviously a hundred times farther away from Earth, aren't they?

Nope.

That simple argument boldly assumes that all stars are intrinsically equally luminous, automatically making the near ones brighter than the far ones. Stars, however, come in a staggering range of luminosities, spanning ten orders of magnitude—ten powers of 10. So the brightest stars are not necessarily the ones closest to Earth. In fact, most of the stars you see in the night sky are of the highly luminous variety, and they lie extraordinarily far away.

If most of the stars we see are highly luminous, then surely those stars are common throughout the galaxy.

Nope again.

High-luminosity stars are the rarest of them all. In any given volume of space, they're outnumbered by the low-luminosity stars a thousand to one. The prodigious energy output of high-luminosity stars is what enables you to see them across such large volumes of space.

Suppose two stars emit light at the same rate (meaning that they have the same luminosity), but one is a hundred times farther from us than the other. We might expect it to be a hundredth as bright. No. That would be too easy. Fact is, the intensity of light dims in proportion to the square of the distance. So in this case, the faraway star looks ten thousand (100^2) times dimmer than the one nearby. The effect of this "inverse-square law" is purely geometric. When starlight spreads in all directions, it dilutes from the growing spherical shell of space through which it moves. The surface area of this sphere increases in proportion to the square of its radius (you may remember the formula: Area = $4\pi r^2$), forcing the light's intensity to diminish by the same proportion.

ALL RIGHT: the stars don't all lie the same distance from us; they aren't all equally luminous; the ones we see are highly unrepresentative. But surely they are stationary in space. For millennia, people understandably thought of stars as "fixed," a concept evident in such influential sources as the Bible ("And God set them in the firmament of the heaven," Genesis 1:17) and Claudius Ptolemy's *Almagest*, published circa A.D. 150, wherein he argues strongly and persuasively for no motion.

To sum up, if you allow the heavenly bodies to move individually, then their distances, measured from Earth upward, must vary. This will force the sizes, brightnesses, and relative separations among the stars to vary too from year to year. But no such variation is apparent. Why? You just didn't wait long enough. Edmond Halley (of comet fame) was the first to figure out that stars moved. In 1718 he compared "modern" star positions with the ones mapped by the second-century B.C. Greek astronomer Hipparchus. Halley trusted the accuracy of Hipparchus's maps, but he also benefited from a baseline of more than eighteen centuries from which to compare the ancient and modern star positions. He promptly noticed that the star Arcturus was not where it once was. The star had indeed moved, but not enough within a single human lifetime to be noticed without the aid of a telescope.

Among all objects in the sky, seven made no pretense of being fixed; they appeared to wander against the starry sky and so were called *planetes*, or "wanderers," by the Greeks. You know all seven (our names for the days of the week can be traced to them): Mercury, Venus, Mars, Jupiter, Saturn, the Sun, and the Moon. Since ancient times, these wanderers were correctly thought to be closer to Earth than were the stars, but each revolving around Earth in the center of it all.

Aristarchus of Samos first proposed a Sun-centered universe in the third century B.C. But back then it was obvious to anybody who paid attention that irrespective of the planets' complicated motions, they and all the background stars revolved around Earth.

If Earth moved we would surely feel it. Common arguments of the day included:

- If Earth rotated on an axis or moved through space, wouldn't clouds in the sky and birds in flight get left far behind? (They aren't.)
- If you jumped vertically, wouldn't you land in a very different spot as Earth traveled swiftly beneath your feet? (You don't.)
- And if Earth moved around the Sun, wouldn't the angle at which we view the stars change continuously, creating a visible shift in the stars' positions on the sky? (It doesn't. At least not visibly.)

The naysayers' evidence was compelling. For the first two cases, the work of Galileo Galilei would later demonstrate that while you are airborne, you, the atmosphere, and everything else around you get carried forward with the rotating, orbiting Earth. For the same reason, if you stand in the aisle of a cruising airplane and jump, you do not catapult backward past the rear seats and get pinned against the lavatory doors. In the third case, there's nothing wrong with the reasoning—except that the stars are so far away you need a powerful telescope to see the seasonal shifts. That effect would not be measured until 1838, by the German astronomer Friedrich Wilhelm Bessel.

The geocentric universe became a pillar of Ptolemy's *Almagest*, and the idea preoccupied scientific, cultural, and religious consciousness until the 1543 publication of *De Revolutionibus*, when Nicolaus Copernicus placed the Sun instead of Earth at the center of the known universe. Fearful that this heretical work would freak out the establishment, Andreas Osiander, a Protestant theologian who oversaw the late stages of the printing, supplied an unauthorized and unsigned preface to the work, in which he pleads:

> I have no doubt that certain learned men, now that the novelty of the hypothesis in this work has been widely reported—for it establishes

that the Earth moves and indeed that the Sun is motionless in the middle of the universe—are extremely shocked. . . . [But it is not] necessary that these hypotheses should be true, nor even probable, but it is sufficient if they merely produce calculations which agree with the observations. (1999, p. 22)

Copernicus himself was not unmindful of the trouble he was about to cause. In the book's dedication, addressed to Pope Paul III, Copernicus notes:

I can well appreciate, Holy Father, that as soon as certain people real-ize that in these books which I have written about the Revolutions of the spheres of the universe I attribute certain motions to the globe of the Earth, they will at once clamor for me to be hooted off the stage with such an opinion. (1999, p. 23)

But soon after the Dutch spectacle maker Hans Lippershey had invented the telescope in 1608, Galileo, using a telescope of his own manufacture, saw Venus going through phases, and four moons that orbited Jupiter and not Earth. These and other obser-vations were nails in the geocentric coffin, making Copernicus's heliocentric universe an increasingly persuasive concept. Once Earth no longer occupied a unique place in the cosmos, the Copernican revolution, based on the principle that we are not spe-cial, had officially begun.

NOW THAT EARTH was in solar orbit, just like its planetary brethren, where did that put the Sun? At the center of the universe? No way. Nobody was going to fall for that one again; it would vio-late the freshly minted Copernican principle. But let's investigate to make sure.

If the solar system were in the center of the universe, then no matter where we looked on the sky we would see approximately the same number of stars. But if the solar system were off to the side somewhere, we would presumably see a great concentration of stars in one direction—the direction of the center of the universe.

By 1785, having tallied stars everywhere on the sky and crudely estimated their distances, the English astronomer Sir William Herschel concluded that the solar system did indeed lie at the center of the cosmos. Slightly more than a century later, the Dutch astronomer Jacobus Cornelius Kapteyn—using the best available methods for calculating distance—sought to verify once and for all the location of the solar system in the galaxy. When seen through a telescope, the band of light called the Milky Way resolves into dense concentrations of stars. Careful tallies of their positions and distances yield similar numbers of stars in every direction along the band itself. Above and below it, the concentration of stars drops symmetrically. No matter which way you look on the sky, the numbers comes out about the same as they do in the opposite direction, 180 degrees away. Kapteyn devoted some 20 years to preparing his sky map, which, sure enough, showed the solar system lying within the central 1 percent of the universe. We weren't in the exact center, but we were close enough to reclaim our rightful place in space.

But the cosmic cruelty continued.

Little did anybody know at the time, especially not Kapteyn, that most sight lines to the Milky Way do not pass all the way through to the end of the universe. The Milky Way is rich in large clouds of gas and dust that absorb the light emitted by objects behind them. When we look in the direction of the Milky Way, more than 99 percent of all stars that should be visible to us are blocked from view by gas clouds within the Milky Way itself. To presume that Earth was near the center of the Milky Way (the then-known universe) was like walking into a large, dense forest and, after a few dozen steps, asserting that you've reached the center simply because you see the same number of trees in every direction.

By 1920—but before the light-absorption problem was well understood—Harlow Shapley, who was to become director of the Harvard College Observatory, studied the spatial layout of globular clusters in the Milky Way. Globular clusters are tight concentrations of as many as a million stars and are seen easily in regions above and below the Milky Way, where the least amount of light is absorbed. Shapley reasoned that these titanic clusters should enable him to pinpoint the center of the universe—a spot that, after all, would surely have the highest concentration of mass and the strongest gravity. Shapley's data showed that the solar system is nowhere close to the center of the globular clusters' distribution, and so is nowhere close to the center of the known universe. Where was this special place he found? Sixty thousand light-years away, in roughly the same direction as—but far beyond—the stars that trace the constellation Sagittarius.

Shapley's distances were too large by more than a factor of 2, but he was right about the center of the system of globular clusters. It coincides with what was later found to be the most powerful source of radio waves in the night sky (radio waves are unattenuated by intervening gas and dust). Astrophysicists eventually identified the site of peak radio emissions as the exact center of the Milky Way, but not until one or two more episodes of seeing-isn't-believing had taken place.

Once again the Copernican principle had triumphed. The solar system was not in the center of the known universe but far out in the suburbs. For sensitive egos, that could still be okay. Surely the vast system of stars and nebulae to which we belong comprised the entire universe. Surely we were where the action was.

Nope.

Most of the nebulae in the night sky are like island universes, as presciently proposed in the eighteenth century by several people, including the Swedish philosopher Emanuel Swedenborg, the English astronomer Thomas Wright, and the German philosopher Immanuel Kant. In *An Original Theory of the Universe* (1750), for instance, Wright speculates on the infinity of space, filled with stellar systems akin to our own Milky Way:

> *We may conclude . . . that as the visible Creation is supposed to be full*
> *of sidereal Systems and planetary Worlds, . . . the endless Immensity*

is an unlimited Plenum of Creations not unlike the known Universe.
. . . That this in all Probability may be the real Case, is in some
Degree made evident by the many cloudy Spots, just perceivable by
us, as far without our starry Regions, in which tho' visibly luminous
Spaces, no one Star or particular constituent Body can possibly be
distinguished; those in all likelyhood may be external Creation, bor-
dering upon the known one, too remote for even our Telescopes to
reach. (p. 177)

Wright's "cloudy Spots" are in fact collections of hundreds of billions of stars, situated far away in space and visible primarily above
and below the Milky Way. The rest of the nebulae turn out to be relatively small, nearby clouds of gas, found mostly within the Milky
Way band.

That the Milky Way is just one of multitudes of galaxies that
comprise the universe was among the most important discoveries
in the history of science, even if it made us feel small again. The
offending astronomer was Edwin Hubble, after whom the *Hubble
Space Telescope* is named. The offending evidence came in the form
of a photographic plate taken on the night of October 5, 1923. The
offending instrument was the Mount Wilson Observatory's 100-
inch telescope, at the time the most powerful in the world. The
offending cosmic object was the Andromeda nebula, one of the
largest on the night sky.

Hubble discovered a highly luminous kind of star within
Andromeda that was already familiar to astronomers from surveys
of stars much closer to home. The distances to the nearby stars
were known, and their brightness varies only with their distance.
By applying the inverse-square law for the brightness of starlight,
Hubble derived a distance to the star in Andromeda, placing the
nebula far beyond any known star within our own stellar system.
Andromeda was actually an entire galaxy, whose fuzz could be
resolved into billions of stars, all situated more than 2 million
light-years away. Not only were we not in the center of things, but
overnight our entire Milky Way galaxy, the last measure of our

self-worth, shrank to an insignificant smudge in a multibillion-smudge universe that was vastly larger than anyone had previously imagined.

ALTHOUGH THE MILKY WAY turned out to be only one of countless galaxies, couldn't we still be at the center of the universe? Just six years after Hubble demoted us, he pooled all the available data on the motions of galaxies. Turns out that nearly all of them recede from the Milky Way, at velocities directly proportional to their distances from us.

Finally we were in the middle of something big: the universe was expanding, and we were its center.

No, we weren't going to be fooled again. Just because it looks as if we're in the center of the cosmos doesn't mean we are. As a matter of fact, a theory of the universe had been waiting in the wings since 1916, when Albert Einstein published his paper on general relativity—the modern theory of gravity. In Einstein's universe, the fabric of space and time warps in the presence of mass. This warping, and the movement of objects in response to it, is what we interpret as the force of gravity. When applied to the cosmos, general relativity allows the space of the universe to expand, carrying its constituent galaxies along for the ride.

A remarkable consequence of this new reality is that the universe looks to all observers in every galaxy as though it expands around them. It's the ultimate illusion of self-importance, where nature fools not only sentient human beings on Earth, but all life-forms that have ever lived in all of space and time.

But surely there is only one cosmos—the one where we live in happy delusion. At the moment, cosmologists have no evidence for more than one universe. But if you extend several well-tested laws of physics to their extremes (or beyond), you can describe the small, dense, hot birth of the universe as a seething foam of tangled space-time that is prone to quantum fluctuations, any one of which could spawn an entire universe of its own. In this gnarly cosmos we might occupy just one universe in a "multiverse" that encompasses countless other universes popping in and out of existence. The idea relegates us to an embarrassingly smaller part of the whole than we ever imagined. What would Pope Paul III think?

OUR PLIGHT PERSISTS, but on ever larger scales. Hubble summarized the issues in his 1936 work *Realm of the Nebulae*, but these words could apply at all stages of our endarkenment:

> *Thus the explorations of space end on a note of uncertainty. . . . We know our immediate neighborhood rather intimately. With increasing distance our knowledge fades, and fades rapidly. Eventually, we reach the dim boundary—the utmost limits of our telescopes. There, we measure shadows, and we search among ghostly errors of measurement for landmarks that are scarcely more substantial. (p. 201)*

What are the lessons to be learned from this journey of the mind? That humans are emotionally fragile, perennially gullible, hopelessly ignorant masters of an insignificantly small speck in the cosmos.

Have a nice day.

FOUR

THE INFORMATION TRAP

Most people assume that the more information you have about something, the better you understand it. Up to a point, that's usually true. When you look at this page from across the room, you can see it's in a book, but you probably can't make out the words. Get close enough, and you'll be able to read the chapter. If you put your nose right up against the page, though, your understanding of the chapter's contents does not improve. You may see more detail, but you'll sacrifice crucial information—whole words, entire sentences, complete paragraphs. The old story about the blind men and the elephant makes the same point: if you stand a few inches away and fixate on the hard, pointed projections, or the long rubbery hose, or the thick, wrinkled posts, or the dangling rope with a tassel on the end that you quickly learn not to pull, you won't be able to tell much about the animal as a whole.

One of the challenges of scientific inquiry is knowing when to step back—and how far back to step—and when to move in close. In some contexts, approximation brings clarity; in others it leads to oversimplification. A raft of complications sometimes points to true complexity and sometimes just clutters up the picture. If you want to know the overall properties of an ensemble of molecules under various states of pressure and temperature, for instance, it's irrelevant and sometimes downright misleading to pay attention to what individual molecules are doing. As we will see in Section 3, a single particle cannot have a temperature, because the very concept of temperature addresses the average motion of all the molecules in the group. In biochemistry, by contrast, you understand next to nothing unless you pay attention to how one molecule interacts with another.

So, when does a measurement, an observation, or simply a map have the right amount of detail?

IN 1967 BENOIT B. MANDELBROT, a mathematician now at IBM's Thomas J. Watson Research Center in Yorktown Heights, New York, and also at Yale University, posed a question in the journal *Science*: "How long is the coast of Britain?" A simple question with a simple answer, you might expect. But the answer is deeper than anyone had imagined.

Explorers and cartographers have been mapping coastlines for centuries. The earliest drawings depict the continents as having crude, funny-looking boundaries; today's high-resolution maps, enabled by satellites, are worlds away in precision. To begin to answer Mandelbrot's question, however, all you need is a handy world atlas and a spool of string. Unwind the string along the perimeter of Britain, from Dunnet Head down to Lizard Point, making sure you go into all the bays and headlands. Then unfurl the string, compare its length to the scale on the map, and voilà! you've measured the island's coastline.

Wanting to spot-check your work, you get hold of a more detailed ordnance survey map, scaled at, say, 2.5 inches to the mile, as opposed to the kind of map that shows all of Britain on a single panel. Now there are inlets and spits and promontories that you'll have to trace with your string; the variations are small, but there are lots of them. You find that the survey map shows the coastline to be longer than the atlas did.

So which measurement is correct? Surely it's the one based on the more detailed map. Yet you could have chosen a map that has even more detail—one that shows every boulder that sits at the base of every cliff. But cartographers usually ignore rocks on a map, unless they're the size of Gibraltar. So, I guess you'll just have to walk the coastline of Britain yourself if you really want to measure it accurately—and you'd better carry a very long string so that you can run it around every nook and cranny. But you'll still be leaving out some pebbles, not to mention the rivulets of water trickling among the grains of sand.

Where does all this end? Each time you measure it, the coastline gets longer and longer. If you take into account the boundaries of molecules, atoms, subatomic particles, will the coastline prove to be infinitely long? Not exactly. Mandelbrot would say "indefinable." Maybe we need the help of another dimension to rethink the problem. Perhaps the concept of one-dimensional length is simply ill-suited for convoluted coastlines.

Playing out Mandelbrot's mental exercise involved a newly synthesized field of mathematics, based on fractional—or fractal (from the Latin *fractus*, "broken")—dimensions rather than the one, two, and three dimensions of classic Euclidean geometry. The ordinary concepts of dimension, Mandelbrot argued, are just too simplistic to characterize the complexity of coastlines. Turns out, fractals are ideal for describing "self-similar" patterns, which look much the same at different scales. Broccoli, ferns, and snowflakes are good examples from the natural world, but only certain computer-generated, indefinitely repeating structures can produce the ideal fractal, in which the shape of the macro object is made up of smaller versions of the same shape or pattern, which are in turn formed from even more miniature versions of the very same thing, and so on indefinitely.

As you descend into a pure fractal, however, even though its components multiply, no new information comes your way—because the pattern continues to look the same. By contrast, if you look deeper and deeper into the human body, you eventually encounter a cell, an enormously complex structure endowed with different attributes and operating under different rules than the ones that hold sway at the macro levels of the body. Crossing the boundary into the cell reveals a new universe of information.

HOW ABOUT EARTH itself? One of the earliest representations of the world, preserved on a 2,600-year-old Babylonian clay tablet, depicts it as a disk encircled by oceans. Fact is, when you stand in the middle of a broad plain (the valley of the Tigris and Euphrates rivers, for instance) and check out the view in every direction, Earth does look like a flat disk.

Noticing a few problems with the concept of a flat Earth, the ancient Greeks—including such thinkers as Pythagoras and Herodotus—pondered the possibility that Earth might be a sphere. In the fourth century B.C., Aristotle, the great systematizer of knowledge, summarized several arguments in support of that view. One of them was based on lunar eclipses. Every now and then, the Moon, as it orbits Earth, intercepts the cone-shaped shadow that Earth casts in space. Across decades of these spectacles, Aristotle noted, Earth's shadow on the Moon was always circular. For that to be true, Earth had to be a sphere, because only spheres cast circular shadows via all light sources, from all angles, and at all times. If Earth were a flat disk, the shadow would sometimes be oval. And some other times, when Earth's edge faced the Sun, the shadow would be a thin line. Only when Earth was face-on to the Sun would its shadow cast a circle.

Given the strength of that one argument, you might think cartographers would have made a spherical model of Earth within the next few centuries. But no. The earliest known terrestrial globe would wait until 1490–92, on the eve of the European ocean voyages of discovery and colonization.

SO, YES, EARTH is a sphere. But the devil, as always, lurks in the details. In Newton's 1687 *Principia*, he proposed that, because spinning spherical objects thrust their substance outward as they rotate, our planet (and the others as well) will be a bit flattened at the poles and a bit bulgy at the equator—a shape known as an oblate spheroid. To test Newton's hypothesis, half a century later, the French Academy of Sciences in Paris sent mathematicians on two expeditions—one to the Arctic Circle and one to the equator—both assigned to measure the length of one degree of latitude on Earth's surface along the same line of longitude. The degree was slightly longer at the Arctic Circle, which could only be true if Earth were a bit flattened. Newton was right.

The faster a planet spins, the greater we expect its equatorial bulge to be. A single day on fast-spinning Jupiter, the most massive planet in the solar system, lasts 10 Earth-hours; Jupiter is 7 percent wider at its equator than at its poles. Our much smaller Earth, with its 24-hour day, is just 0.3 percent wider at the equator—27 miles on a diameter of just under 8,000 miles. That's hardly anything.

One fascinating consequence of this mild oblateness is that if you stand at sea level anywhere on the equator, you'll be farther from Earth's center than you'd be nearly anywhere else on Earth. And if you really want to do things right, climb Mount Chimborazo in central Ecuador, close to the equator. Chimborazo's summit is four miles above sea level, but more important, it sits 1.33 miles farther from Earth's center than does the summit of Mount Everest.

SATELLITES HAVE MANAGED to complicate matters further. In 1958 the small Earth orbiter *Vanguard 1* sent back the news that the equatorial bulge south of the equator was slightly bulgier than the bulge north of the equator. Not only that, sea level at the South Pole turned out to be a tad closer to the center of Earth than sea level at the North Pole. In other words, the planet's a pear.

Next up is the disconcerting fact that Earth is not rigid. Its surface rises and falls daily as the oceans slosh in and out of the continental shelves, pulled by the Moon and, to a lesser extent, by the

Sun. Tidal forces distort the waters of the world, making their surface oval. A well-known phenomenon. But tidal forces stretch the solid earth as well, and so the equatorial radius fluctuates daily and monthly, in tandem with the oceanic tides and the phases of the Moon.

So Earth's a pearlike, oblate-spheroidal hula hoop.

Will the refinements never end? Perhaps not. Fast forward to 2002. A U.S.-German space mission named GRACE (Gravity Recovery and Climate Experiment) sent up a pair of satellites to map Earth's geoid, which is the shape Earth would have if sea level were unaffected by ocean currents, tides, or weather—in other words, a hypothetical surface where the force of gravity is perpendicular to every mapped point. Thus, the geoid embodies the truly horizontal, fully accounting for all the variations in Earth shape and subsurface density of matter. Carpenters, land surveyors, and aqueduct engineers will have no choice but to obey.

ORBITS ARE ANOTHER category of problematic shape. They're not one-dimensional, nor merely two- or three-dimensional. Orbits are multidimensional, unfolding in both space and time. Aristotle advanced the idea that Earth, the Sun, and the stars were locked in place, attached to crystalline spheres. It was the spheres that rotated, and their orbits traced—what else?—perfect circles. To Aristotle and nearly all the ancients, Earth lay at the center of all this activity.

Nicolaus Copernicus disagreed. In his 1543 magnum opus, *De Revolutionibus,* he placed the Sun in the middle of the cosmos. Copernicus nonetheless maintained perfect circular orbits, unaware of their mismatch with reality. Half a century later, Johannes Kepler put matters right with his three laws of planetary motion—the first predictive equations in the history of science— one of which showed that the orbits are not circles but ovals of varying elongation.

We have only just begun.

Consider the Earth-Moon system. The two bodies orbit their common center of mass, their barycenter, which lies roughly 1,000 miles below the spot on Earth's surface closest to the Moon at any given moment. So instead of the planets themselves, it's actually their planet-moon barycenters that trace the Keplerian elliptical orbits around the Sun. So now what's Earth's trajectory? A series of loop-the-loops—thirteen of them in a year, one for each cycle of lunar phases—rolled together with an ellipse.

Meanwhile, not only do the Moon and Earth tug on each other, but all the other planets (and their moons) tug on them too. Everybody's tugging on everybody else. As you might suspect, it's a complicated mess, and will be described further in Section 3. Plus, each time the Earth-Moon system takes a trip around the Sun, the orientation of the ellipse shifts slightly, not to mention that the Moon is spiraling away from Earth at a rate of one or two inches per year and that some orbits in the solar system are chaotic.

All told, this ballet of the solar system, choreographed by the forces of gravity, is a performance only a computer can know and love. We've come a long way from single, isolated bodies tracing pure circles in space.

THE COURSE OF a scientific discipline gets shaped in different ways, depending on whether theories lead data or data lead theories. A theory tells you what to look for, and you either find it or you don't. If you find it, you move on to the next open question. If you have no theory but you wield tools of measurement, you'll start collecting as much data as you can and hope that patterns emerge. But until you arrive at an overview, you're mostly poking around in the dark.

Nevertheless, one would be misguided to declare that Copernicus was wrong simply because his orbits were the wrong shape. His deeper concept—that planets orbit the Sun—is what mattered most. From then on, astrophysicists have continually refined the model by looking closer and closer. Copernicus may not have been in the right ballpark, but he was surely on the right side of town. So, perhaps, the question still remains: When do you move closer and when do you take a step back?

NOW IMAGINE YOU'RE strolling along a boulevard on a crisp autumn day. A block ahead of you is a silver-haired gentleman wearing a dark blue suit. It's unlikely you'll be able to see the jewelry on his left hand. If you quicken your pace and get within 30 feet of him, you might notice he's wearing a ring, but you won't see its crimson stone or the designs on its surface. Sidle up close with a magnifying glass and—if he doesn't alert the authorities—you'll learn the name of the school, the degree he earned, the year he graduated, and possibly the school emblem. In this case, you've correctly assumed that a closer look would tell you more.

Next, imagine you're gazing at a late-nineteenth-century French pointillist painting. If you stand 10 feet away, you might see men in tophats, women in long skirts and bustles, children, pets, shimmering water. Up close, you'll just see tens of thousands of dashes, dots, and streaks of color. With your nose on the canvas you'll be able to appreciate the complexity and obsessiveness of the technique, but only from afar will the painting resolve into the representation of a scene. It's the opposite of your experience with the ringed gentleman on the boulevard: the closer you look at a pointillist masterpiece, the more the details disintegrate, leaving you wishing you had kept your distance.

Which way best captures how nature reveals itself to us? Both, really. Almost every time scientists look more closely at a phenomenon, or at some inhabitant of the cosmos, whether animal, vegetable, or star, they must assess whether the broad picture—the one you get when you step back a few feet—is more useful or less useful than the close-up. But there's a third way, a kind of hybrid of the two, in which looking closer gives you more data, but the extra data leave you extra baffled. The urge to pull back is strong, but so, too, is the urge to push ahead. For every hypothesis that gets confirmed by more detailed data, ten others will have to be modified or discarded altogether because they no longer fit the model. And years or decades may pass before the half-dozen new insights based on those data are even formulated. Case in point: the multitudinous rings and ringlets of the planet Saturn.

EARTH IS A FASCINATING PLACE to live and work. But before Galileo first looked up with a telescope in 1609, nobody had any awareness or understanding of the surface, composition, or climate of any other place in the cosmos. In 1610 Galileo noticed something odd about Saturn; because the resolution of his telescope was poor, however, the planet looked to him as if it had two companions, one to its left and one to its right. Galileo formulated his observation in an anagram,

smaismrmilmepoetaleumibunenugttauiras

designed to ensure that no one else could snatch prior credit for his radical and as-yet-unpublished discovery. When sorted out and translated from the Latin, the anagram becomes: "I have observed the highest planet to be triple-bodied." As the years went by, Galileo continued to monitor Saturn's companions. At one stage they looked like ears; at another stage they vanished completely.

In 1656 the Dutch physicist Christiaan Huygens viewed Saturn through a telescope of much higher resolution than Galileo's, built for the express purpose of scrutinizing the planet. He became the first to interpret Saturn's earlike companions as a simple, flat ring. As Galileo had done half a century earlier, Huygens wrote down his groundbreaking but still preliminary finding in the form of an anagram. Within three years, in his book *Systema Saturnium,* Huygens went public with his proposal.

Twenty years later Giovanni Cassini, the director of the Paris Observatory, pointed out that there were two rings, separated by a gap that came to be known as the Cassini division. And nearly two centuries later, the Scottish physicist James Clerk Maxwell won a prestigious prize for showing that Saturn's rings are not solid, but made up instead of numerous small particles in their own orbits.

By the end of the twentieth century, observers had identified seven distinct rings, lettered A through G. Not only that, the rings themselves turn out to be made up of thousands upon thousands of bands and ringlets.

So much for the "ear theory" of Saturn's rings.

SEVERAL SATURN FLYBYS took place in the twentieth century: *Pioneer 11* in 1979, *Voyager 1* in 1980, and *Voyager 2* in 1981. Those relatively close inspections all yielded evidence that the ring system is more complex and more puzzling than anyone had imagined. For one thing, the particles in some of the rings corral into narrow bands by the so-called shepherd moons: teeny satellites that orbit near and within the rings. The gravitational forces of the shepherd moons tug the ring particles in different directions, sustaining numerous gaps among the rings.

Density waves, orbital resonances, and other quirks of gravitation in multiple-particle systems give rise to passing features within and among the rings. Ghostly, shifting "spokes" in Saturn's B ring, for instance—recorded by the *Voyager* space probes and presumed to be caused by the planet's magnetic field—have mysteriously vanished from close-up views supplied by the *Cassini* spacecraft, sending images from Saturnian orbit.

What kind of stuff are Saturn's rings made of? Water ice, for the most part—though there's also some dirt mixed in, whose chemical makeup is similar to one of the planet's larger moons. The cosmochemistry of the environment suggests that Saturn might once have had several such moons. Those that went AWOL may have orbited too close for comfort to the giant planet and gotten ripped apart by Saturn's tidal forces.

Saturn, by the way, is not the only planet with a ring system. Close-up views of Jupiter, Uranus, and Neptune—the rest of the big four gas giants in our solar system—show that each planet bears a ring system of its own. The Jovian, Uranian, and Neptunian rings weren't discovered until the late 1970s and early 1980s, because, unlike Saturn's majestic ring system, they're made largely of dark, unreflective substances such as rocks or dust grains.

THE SPACE NEAR a planet can be dangerous if you're not a dense, rigid object. As we will see in Section 2, many comets and some asteroids, for instance, resemble piles of rubble, and they swing near planets at their peril. The magic distance, within which a planet's tidal force exceeds the gravity holding together that kind of vagabond, is called the Roche limit—discovered by the nineteenth-century French astronomer Édouard Albert Roche. Wander inside the Roche limit, and you'll get torn apart; your disassembled bits and pieces will then scatter into their own orbits and eventually spread out into a broad, flat, circular ring.

I recently received some upsetting news about Saturn from a colleague who studies ring systems. He noted with sadness that the orbits of their constituent particles are unstable, and so the particles will all be gone in an astrophysical blink of an eye: 100 million years or so. My favorite planet, shorn of what makes it my favorite planet! Turns out, fortunately, that the steady and essentially unending accretion of interplanetary and intermoon particles may replenish the rings. The ring system—like the skin on your face—may persist, even if its constituent particles do not.

Other news has come to Earth via *Cassini*'s close-up pictures of Saturn's rings. What kind of news? "Mind-boggling" and "startling," to quote Carolyn C. Porco, the leader of the mission's imaging team and a specialist in planetary rings at the Space Science Institute in

Boulder, Colorado. Here and there in all those rings are features neither expected nor, at present, explainable: scalloped ringlets with extremely sharp edges, particles coalescing in clumps, the pristine iciness of the A and B rings compared with the dirtiness of the Cassini division between them. All these new data will keep Porco and her colleagues busy for years to come, perhaps wistfully recalling the clearer, simpler view from afar.

FIVE

STICK-IN-THE-MUD SCIENCE

For a century or two, various blends of high technology and clever thinking have driven cosmic discovery. But suppose you have no technology. Suppose all you have in your backyard laboratory is a stick. What can you learn? Plenty.

With patience and careful measurement, you and your stick can glean an outrageous amount of information about our place in the cosmos. It doesn't matter what the stick is made of. And it doesn't matter what color it is. The stick just has to be straight. Hammer the stick firmly into the ground where you have a clear view of the horizon. Since you're going low-tech, you might as well use a rock for a hammer. Make sure the stick isn't floppy and that it stands up straight.

Your caveman laboratory is now ready.

On a clear morning, track the length of the stick's shadow as the Sun rises, crosses the sky, and finally sets. The shadow will start long, get shorter and shorter until the Sun reaches its highest point in the sky, and finally lengthen again until sunset. Collecting data for this experiment is about as exciting as watching the hour hand move on a clock. But since you have no technology, not much else competes for your attention. Notice that when the shadow is shortest, half the day has passed. At that moment—called local noon—the shadow points due north or due south, depending which side of the equator you're on.

You've just made a rudimentary sundial. And if you want to sound erudite, you can now call the stick a gnomon (I still prefer "stick"). Note that in the Northern Hemisphere, where civilization began, the stick's shadow will revolve clockwise around the base of the stick as the Sun moves across the sky. Indeed, that's why the hands of a clock turn "clockwise" in the first place.

If you have enough patience and cloudless skies to repeat the exercise 365 times in a row, you will notice that the Sun doesn't rise from day to day at the same spot on the horizon. And on two days a year the shadow of the stick at sunrise points exactly opposite the shadow of the stick at sunset. When that happens, the Sun rises due cast, sets due west, and daylight lasts as long as night. Those two days are the spring and fall equinoxes (from the Latin for "equal night"). On all other days of the year the Sun rises and sets elsewhere along the horizon. So the person who invented the adage "the Sun always raises in the east and sets in the west" simply never paid attention to the sky.

If you're in the Northern Hemisphere while tracking the rise and set points for the Sun, you'll see that those spots creep north of the east-west line after the spring equinox, eventually stop, and then creep south for a while. After they cross the east-west line again, the southward creeping eventually slows down, stops, and gives way to the northward creeping once again. The entire cycle repeats annually.

All the while, the Sun's trajectory is changing. On the summer solstice (Latin for "stationary Sun"), the Sun rises and sets at its northernmost point along the horizon, tracing its highest path across the sky. That makes the solstice the year's longest day, and the stick's noontime shadow on that day the shortest. When the Sun rises and sets at its southernmost point along the horizon, its trajectory across the sky is the lowest, creating the year's longest noontime shadow. What else to call that day but the winter solstice?

For 60 percent of Earth's surface and about 75 percent of its human inhabitants, the Sun is never, ever directly overhead. For the rest of our planet, a 3,200-mile-wide belt centered on the equator, the Sun climbs to the zenith only two days a year (okay, just one day a year if you're smack on the Tropic of Cancer or the Tropic of Capricorn). I'd bet the same person who professed to know where the Sun rises and sets on the horizon also started the adage "the Sun is directly overhead at high noon."

So far, with a single stick and profound patience, you have identified the cardinal points on the compass and the four days of the year that mark the change of seasons. Now you need to invent some way to time the interval between one day's local noon and the next. An expensive chronometer would help here, but one or more well-made hourglasses will also do just fine. Either timer will enable you to determine, with great accuracy, how long it takes for the Sun to revolve around Earth: the solar day. Averaged over the entire year, that time interval equals 24 hours, exactly. Although this doesn't include the leap-second added now and then to account for the slowing of Earth's rotation by the Moon's gravitational tug on Earth's oceans.

Back to you and your stick. We're not done yet. Establish a line of sight from its tip to a spot on the sky, and use your trusty timer to mark the moment a familiar star from a familiar constellation passes by. Then, still using your timer, record how long it takes for the star to realign with the stick from one night to the next. That interval, the sidereal day, lasts 23 hours, 56 minutes, and 4 seconds. The almost-four-minute mismatch between the sidereal and solar days forces the Sun to migrate across the patterns of background stars, creating the impression that the Sun visits the stars in one constellation after another throughout the year.

Of course, you can't see stars in the daytime—other than the Sun. But the ones visible near the horizon just after sunset or just before sunrise flank the Sun's position on the sky, and so a sharp observer with a good memory for star patterns can figure out what patterns lie behind the Sun itself.

Once again taking advantage of your timing device, you can try something different with your stick in the ground. Each day for an entire year, mark where the tip of the stick's shadow falls at noon, as indicated by your timer. Turns out that each day's mark will fall in a different spot, and by the end of the year you will have traced a figure eight, known to the erudite as an "analemma."

Why? Earth tilts on its axis by 23.5 degrees from the plane of the solar system. This tilt not only gives rise to the familiar seasons and the wide-ranging daily path of the Sun across the sky, it's also the dominant cause of the figure eight that emerges as the Sun migrates back and forth across the celestial equator throughout the year. Moreover, Earth's orbit about the Sun is not a perfect circle. According to Kepler's laws of planetary motion, its orbital speed must vary, increasing as we near the Sun and slowing down as we recede. Because the rate of Earth's rotation remains rock-steady, something has to give: the Sun does not always reach its highest point on the sky at "clock noon." Although the shift is slow from day to day, the Sun gets there as much as 14 minutes late at certain times of year. At other times it's as much as 16 minutes early. On only four days a year—corresponding to the top, the bottom, and the middle crossing of the figure eight—is clock time equal to Sun time. As it happens, the days fall on or about April 15 (no relation to taxes), June 14 (no relation to flags), September 2 (no relation to labor), and December 25 (no relation to Jesus).

Next up, clone yourself and your stick and send your twin due south to a prechosen spot far beyond your horizon. Agree in advance that you will both measure the length of your stick shadows at the same time on the same day. If the shadows are the same length, you live on a flat or a supergigantic Earth. If the shadows have different lengths, you can use simple geometry to calculate Earth's circumference.

The astronomer and mathematician Eratosthenes of Cyrene (276–194 B.C.) did just that. He compared shadow lengths at noon from two Egyptian cities—Syene (now called Aswan) and Alexandria, which he overestimated to be 5,000 stadia apart. Eratosthenes' answer for Earth's circumference was within 15 percent of the correct value. The word "geometry," in fact, comes from the Greek for "earth measurement."

Although you've now been occupied with sticks and stones for several years, the next experiment will take only about a minute. Pound your stick into the ground at an angle other than vertical, so that it resembles a typical stick in the mud. Tie a stone to the end of a thin string and dangle it from the stick's tip. Now you've got a pendulum. Measure the length of the string and then tap the bob to set the pendulum in motion. Count how many times the bob swings in 60 seconds.

The number, you'll find, depends very little on the width of the pendulum's arc, and not at all on the mass of the bob. The only things that matter are the length of the string and what planet you're on. Working with a relatively simple equation, you can deduce the acceleration of gravity on Earth's surface, which is a direct measure of your weight. On the Moon, with only one-sixth the gravity of Earth, the same pendulum will move much more slowly, executing fewer swings per minute.

There's no better way to take the pulse of a planet.

UNTIL NOW YOUR stick has offered no proof that Earth itself rotates—only that the Sun and the nighttime stars revolve at regular, predictable intervals. For the next experiment, find a stick more than 10 yards long and, once again, pound it into the ground at a tilt. Tie a heavy stone to the end of a long, thin string and dangle it from the tip. Now, just like last time, set it in motion. The long, thin string and the heavy bob will enable the pendulum to swing unencumbered for hours and hours and hours.

If you carefully track the direction the pendulum swings, and if you're extremely patient, you will notice that the plane of its swing slowly rotates. The most pedagogically useful place to do this experiment is at the geographic North (or, equivalently, South) Pole. At the Poles, the plane of the pendulum's swing makes one full rotation in 24 hours—a simple measure of the direction and rotational speed of the earth beneath it. For all other positions on Earth, except along the equator, the plane still turns, but more and more slowly as you move from the Poles toward the equator. At the equator the plane of the pendulum does not move at all. Not only does this experiment demonstrate that it's Earth, not the Sun, that moves, but with the help of a little trigonometry you can also turn the question around and use the time needed for one rotation of the pendulum's plane to determine your geographic latitude on our planet.

The first person to do this was Jean-Bernard-Léon Foucault, a French physicist who surely conducted the last of the truly cheap laboratory experiments. In 1851 he invited his colleagues to "come and see the Earth turn" at the Pantheon in Paris. Today a Foucault pendulum sways in practically every science and technology museum in the world.

Given all that one can learn from a simple stick in the ground, what are we to make of the world's famous prehistoric observatories? From Europe and Asia to Africa and Latin America, a survey of ancient cultures turns up countless stone monuments that served as low-tech astronomy centers, although it's likely they also doubled as places of worship or embodied other deeply cultural meanings.

On the morning of the summer solstice at Stonehenge, for instance, several of the stones in its concentric circles align precisely with sunrise. Certain other stones align with the extreme rising and setting points of the Moon. Begun in about 3100 B.C. and altered during the next two millennia, Stonehenge incorporates outsize monoliths quarried far from its site on Salisbury Plain in southern England. Eighty or so bluestone pillars, each weighing several tons, came from the Preseli Mountains, roughly 240 miles away. The so-called sarsen stones, each weighing as much as 50 tons, came from Marlborough Downs, 20 miles away.

Much has been written about the significance of Stonehenge. Historians and casual observers alike are impressed by the astronomical knowledge of these ancient people, as well as by their ability to transport such obdurate materials such long distances. Some fantasy-prone observers are so impressed that they even credit extraterrestrial intervention at the time of construction.

Why the ancient civilizations who built the place did not use the easier, nearby rocks remains a mystery. But the skills and knowledge on display at Stonehenge are not. The major phases of construction took a total of a few hundred years. Perhaps the preplanning took another hundred or so. You can build anything in half a millennium—I don't care how far you choose to drag your bricks. Furthermore, the astronomy embodied in Stonehenge is not fundamentally deeper than what can be discovered with a stick in the ground.

Perhaps these ancient observatories perennially impress modern people because modern people have no idea how the Sun, Moon, or stars move. We are too busy watching evening television to care what's going on in the sky. To us, a simple rock alignment based on cosmic patterns looks like an Einsteinian feat. But a truly mysterious civilization would be one that made no cultural or architectural reference to the sky at all.

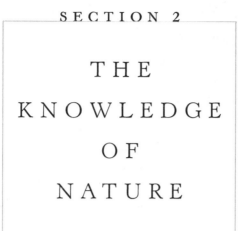

SECTION 2

THE
KNOWLEDGE
OF
NATURE

THE CHALLENGES OF DISCOVERING
THE CONTENTS OF THE COSMOS

JOURNEY FROM THE CENTER
OF THE SUN

During our everyday lives we don't often stop to think about the journey of a ray of light from the core of the Sun, where it's made, all the way to Earth's surface, where it might slam into somebody's buttocks on a sandy beach. The easy part is the ray's 500-second speed-of-light jaunt from the Sun to Earth, through the void of interplanetary space. The hard part is the light's million-year adventure to get from the Sun's center to its surface.

In the cores of stars, beginning at about 10-million degrees Kelvin, but for the Sun, at 15-million degrees, hydrogen nuclei, long denuded of their lone electron, reach high enough speeds to over-come their natural repulsion and collide. Energy is created out of matter as thermonuclear fusion makes a single helium (He) nucleus out of four hydrogen (H) nuclei. Omitting intermediate steps, the Sun simply says:

$$4H \rightarrow He + energy$$
And there is light.

Every time a helium nucleus gets created, particles of light called photons get made. And they pack enough punch to be gamma rays, a form of light with the highest energy for which we have a classifica-tion. Born moving at the speed of light (186,282 miles per second), the gamma-ray photons unwittingly begin their trek out of the Sun.

..n undisturbed photon will always move in a straight line. But if something gets in its way, the photon will either be scattered or absorbed and re-emitted. Each fate can result in the photon being cast in a different direction with a different energy. Given the density of matter in the Sun, the photon's average straight-line trip lasts for less than one thirty-billionth of a second (a thirtieth of a nanosecond)—just long enough for the photon to travel about one centimeter before interacting with a free electron or an atom.

The new travel path after each interaction can be outward, sideways, or even backward. How then does an aimlessly wandering photon ever manage to leave the Sun? A clue lies in what would happen to a fully inebriated person who takes steps in random directions from a street corner lamppost. Curiously, the odds are that the drunkard will not return to the lamppost. If the steps are indeed random, distance from the lamppost will slowly accumulate.

While you cannot predict exactly how far from the lamppost any particular drunk person will be after a selected number of steps, you can reliably predict the average distance if you managed to convince a large number of drunken subjects to randomly walk for you in an experiment. Your data would show that on average, distance from the lamppost increased in proportion to the square root of the total number of paces taken. For example, if each person took 100 steps in random directions, then the average distance from the lamppost would have been a mere 10 steps. If 900 steps were taken, the average distance would have grown to only 30 steps.

With a step size of one centimeter, a photon must execute nearly 5 sextillion steps to "random walk" the 70-billion centimeters from the Sun's center to its surface. The total linear distance traveled would span about 5,000 light-years. At the speed of light, a photon would, of course, take 5,000 years to journey that far. But when computed with a more realistic model of the Sun's profile—taking into account, for example, that about 90 percent of the Sun's mass resides within only half its radius because the gaseous Sun compresses under its own weight—and adding travel time lost during the pit stop between photon absorption and re-emission, the total trip lasts about a million years. If a photon had a clear path from the Sun's center to its surface, its journey would instead last all of 2.3 seconds.

As early as the 1920s, we had some idea that a photon might meet some major resistance getting out of the Sun. Credit the colorful British astrophysicist Sir Arthur Stanley Eddington for endowing the study of stellar structure with enough of a foundation in physics to offer insight into the problem. In 1926 he wrote *The Internal Constitution of the Stars*, which he published immediately after the new branch of physics called quantum mechanics was discovered, but nearly 12 years before thermonuclear fusion was officially credited as the energy source for the Sun. Eddington's glib musings from the introductory chapter correctly capture some of the spirit, if not the detail, of an aether wave's (photon's) tortured journey:

> *The inside of a star is a hurly-burly of atoms, electrons and aether waves. We have to call to aid the most recent discoveries of atomic physics to follow the intricacies of the dance. . . . Try to picture the tumult! Dishevelled atoms tear along at 50 miles a second with only a few tatters left of their elaborate cloaks of electrons torn from them in the scrimmage. The lost electrons are speeding a hundred times faster to find new resting-places. Look out! A thousand narrow shaves happen to the electron in [one ten-billionth] of a second. . . . Then . . . the electron is fairly caught and attached to the atom, and its career of freedom is at an end. But only for an instant. Barely has the atom arranged the new scalp on its girdle when a quantum of aether waves runs into it. With a great explosion the electron is off again for further adventures.* (p. 19)

Eddington's enthusiasm for his subject continues as he identifies aether waves as the only component of the Sun on the move:

> *As we watch the scene we ask ourselves, can this be the stately drama of stellar evolution? It is more like the jolly crockery-smashing turn of a music-hall. The knockabout comedy of atomic physics is not very considerate towards our aesthetic ideals. . . . The atoms and electrons*

for all their hurry never get anywhere; they only change places. The
aether waves are the only part of the population which do actually
accomplish something; although apparently darting about in all
directions without purpose they do in spite of themselves make a slow
general progress outwards. (pp. 19–20)

In the outer one-fourth of the Sun's radius, energy moves primarily through turbulent convection, which is a process not unlike
what happens in a pot of boiling chicken soup (or a pot of boiling
anything). Whole blobs of hot material rise while other blobs of
cooler material sink. Unbeknownst to our hardworking photons,
their residential blob can swiftly sink tens of thousands of kilometers back into the Sun, thus undoing possibly thousands of years of
random walking. Of course the reverse is also true—convection can
swiftly bring random-walking photons near the surface, thus
enhancing their chances of escape.

But the tale of our gamma ray's journey is still not fully told.
From the Sun's 15-million-degree Kelvin center to its 6,000-degree
surface, the temperature drops at an average rate of about one
one-hundredth of a degree per meter. For every absorption and
re-emission, the high-energy gamma-ray photons tend to give birth
to multiple lower-energy photons at the expense of their own existence. Such altruistic acts continue down the spectrum of light
from gamma rays to x-rays to ultraviolet to visible and to the
infrared. The energy from a single gamma-ray photon is sufficient
to beget a thousand x-ray photons, each of which will ultimately
beget a thousand visible-light photons. In other words, a single
gamma ray can easily spawn over a million visible and infrared
photons by the time the random walk reaches the Sun's surface.

Only one out of every half-billion photons that emerge from the
Sun actually heads toward Earth. I know it sounds meager, but at
our size and distance from the Sun it totals Earth's rightful share.
The rest of the photons head everywhere else.

The Sun's gaseous "surface" is, by the way, defined by the layer
where our randomly walking photons take their last step before

escaping to interplanetary space. Only from such a layer can light reach your eye along an unimpeded line of sight, which allows you to assess meaningful solar dimensions. In general, light with longer wavelengths emerges from within deeper layers of the Sun than light of shorter wavelengths. For example, the Sun's diameter is slightly smaller when measured using infrared than when measured with visible light. Whether or not textbooks tell you, their listed values for the Sun's diameter typically assume you seek dimensions obtained using visible light.

Not all the energy of our fecund gamma rays became lower-energy photons. A portion of the energy drives the large-scale turbulent convection, which in turn drives pressure waves that ring the Sun the way a clanger rings a bell. Careful and precise measurements of the Sun's spectrum, when monitored continuously, reveal tiny oscillations that can be interpreted in much the same way that geoseismologists interpret subsurface sound waves induced by earthquakes. The Sun's vibration pattern is extraordinarily complex because many oscillating modes operate simultaneously. The greatest challenges among helioseismologists lie in decomposing the oscillations into their basic parts, and thus deducing the size and structure of the internal features that cause them. A similar "analysis" of your voice would take place if you screamed into an open piano. Your vocal sound waves would induce vibrations of the piano strings that shared the same assortment frequencies that comprise your voice.

A coordinated project to study solar oscillating phenomena was carried out by GONG (yet another cute acronym), the Global Oscillation Network Group. Specially outfitted solar observatories that span the world's time zones (in Hawaii, California, Chile, the Canary Islands, India, and Australia) allowed solar oscillations to be monitored continuously. Their long-anticipated results supported most current notions of stellar structure. In particular, that energy moves by randomly walking photons in the Sun's inner layers and then by large-scale turbulent convection in its outer layers. Yes, some discoveries are great simply because they confirm what you had suspected all along.

Heroic adventures through the Sun are best taken by photons and not by any other form of energy or matter. If any of us were to go on the same trip then we would, of course, be crushed to death, vaporized, and have every single electron stripped from our body's atoms. Aside from these setbacks, I imagine one could easily sell tickets for such a voyage. For me, though, I am content just knowing the story. When I sunbathe, I do it with full respect for the journey made by all photons that hit my body, no matter where on my anatomy they strike.

SEVEN

PLANET PARADE

In the study of the cosmos, it's hard to come up with a better tale than the centuries-long history of attempts to understand the planets—those sky wanderers that make their rounds against the backdrop of stars. Of the eight objects in our solar system that are indisputably planets, five are readily visible to the unaided eye and were known to the ancients, as well as observant troglodytes. Each of the five—Mercury, Venus, Mars, Jupiter, and Saturn—was endowed with the personality of the god for which it was named. For example, Mercury, which moves the fastest against the background stars, was named for the Roman messenger god—the fellow usually depicted with small and aerodynamically useless wings on his heels or his hat. And Mars, the only one of the classic wanderers (the Greek word *planete* means "wanderer") with a reddish hue, was named for the Roman god of war and bloodshed. Earth, of course, is also visible to the unaided eye. Just look down. But terra firma was not identified as one of the gang of planets until after 1543, when Nicolaus Copernicus advanced his Sun-centered model of the universe.

To the telescopically challenged, the planets were, and are, just points of light that happen to move across the sky. Not until the seventeenth century, with the proliferation of telescopes, did astronomers discover that planets were orbs. Not until the twentieth century were the planets scrutinized at close range with space probes. And not until later in the twenty-first century will people be likely to visit them.

Humanity had its first telescopic encounter with the celestial wanderers during the winter of 1609–10. After merely hearing of the 1608 Dutch invention, Galileo Galilei manufactured an excellent telescope of his own design, through which he saw the planets as orbs, perhaps even other worlds. One of them, brilliant Venus, went through phases just like the Moon's: crescent Venus, gibbous Venus, full Venus. Another planet, Jupiter, had moons all of its own, and Galileo discovered the four largest: Ganymede, Callisto, Io, and Europa, all named for assorted characters in the life and times of Jupiter's Greek counterpart, Zeus.

The simplest way to explain the phases of Venus, as well as other features of its motion on the sky, was to assert that the planets revolve around the Sun, not Earth. Indeed, Galileo's observations strongly supported the universe as envisioned and theorized by Copernicus.

Jupiter's moons took the Copernican universe a step further: although Galileo's 20-power telescope could not resolve the moons into anything larger than pinpoints of light, no one had ever seen a celestial object revolve around anything other than Earth. An honest, simple observation of the cosmos, except that the Roman Catholic Church and "common" sense would have none of it. Galileo discovered with his telescope a contradiction to the dogma that Earth occupied the central position in the cosmos—the spot around which all objects revolve. Galileo reported his persuasive findings in early 1610, in a short but seminal work he titled *Sidereus Nuncius* ("the Starry Messenger").

ONCE THE COPERNICAN model became widely accepted, the arrangement of the heavens could legitimately be called a *solar* system, and Earth could take its proper place as one among six known planets. Nobody imagined there could be more than six. Not even the English astronomer Sir William Herschel, who discovered a seventh in 1781.

Actually, the credit for the first recorded sighting of the seventh planet goes to the English astronomer John Flamsteed, the first British Astronomer Royal. But in 1690, when Flamsteed noted the object, he didn't see it move. He assumed it was just another star in the sky, and named it 34 Tauri. When Herschel saw Flamsteed's "star" drift against the background stars, he announced—operating under the unwitting assumption that planets were not on the list of things one might discover—that he had discovered a comet. Comets, after all, were known to move and to be discoverable. Herschel planned to call the newfound object Georgium Sidus ("Star of George"), after his benefactor, King George III of England. If the astronomical community had respected these wishes, the roster of our solar system would now include Mercury, Venus, Earth, Mars, Jupiter, Saturn, and George. In a blow to sycophancy the object was ultimately called Uranus, in keeping with its classically named brethren—though some French and American astronomers kept calling it "Herschel's planet" until 1850, several years after the eighth planet, Neptune, was discovered.

Over time, telescopes kept getting bigger and sharper, but the detail that astronomers could discern on the planets did not much improve. Because every telescope, no matter the size, viewed the planets through Earth's turbulent atmosphere, the best pictures were still a bit fuzzy. But that didn't keep intrepid observers from discovering things like Jupiter's Great Red Spot, Saturn's rings, Martian polar ice caps, and dozens of planetary moons. Still, our knowledge of the planets was meager, and where ignorance lurks, so too do the frontiers of discovery and imagination.

CONSIDER THE CASE of Percival Lowell, the highly imaginative and wealthy American businessman and astronomer, whose endeavors took place at the end of the nineteenth century and the early years of the twentieth. Lowell's name is forever linked with the "canals" of Mars, the "spokes" of Venus, the search for Planet

X, and of course the Lowell Observatory in Flagstaff, Arizona. Like so many investigators around the world, Lowell picked up on the late-nineteenth-century proposition by the Italian astronomer Giovanni Schiaparelli that linear markings visible on the Martian surface were *canali*.

The problem was that the word means "channels," but Lowell chose to translate the word badly as "canals" because the markings were thought to be similar in scale to the major public-works projects on Earth. Lowell's imagination ran amok, and he dedicated himself to the observation and mapping of the Red Planet's network of waterways, surely (or so he fervently believed) constructed by advanced Martians. He believed that the Martian cities, having exhausted their local water supply, needed to dig canals to transport water from the planet's well-known polar ice caps to the more populous equatorial zones. The story was appealing, and it helped generate plenty of vivid writing.

Lowell was also fascinated by Venus, whose ever-present and highly reflective clouds make it one of the brightest objects in the night sky. Venus orbits relatively near the Sun, so as soon as the Sun sets—or just before the Sun rises—there's Venus, hanging gloriously in the twilight. And because the twilight sky can be quite colorful, there's no end of 9-1-1 calls reporting a glowing, light-adorned UFO hovering on the horizon.

Lowell maintained that Venus sported a network of massive, mostly radial spokes (more *canali*) emanating from a central hub. The spokes he saw remained a puzzle. In fact nobody could ever confirm what he saw on either Mars or Venus. This didn't much bother other astronomers because everyone knew that Lowell's mountaintop observatory was one of the finest in the world. So if you weren't seeing Martian activity the way Percival was, it was surely because your telescope and your mountain were not as good as his.

Of course, even after telescopes got better, nobody could duplicate Lowell's findings. And the episode is today remembered as one where the urge to believe undermined the need to obtain accurate and responsible data. And curiously, it was not until the twenty-first century that anybody could explain what was going on at the Lowell Observatory.

An optometrist from Saint Paul, Minnesota, named Sherman Schultz wrote a letter in response to an article in the July 2002 issue of *Sky and Telescope* magazine. Schultz pointed out that the optical setup Lowell preferred for viewing the Venutian surface was similar to the gizmo used to examine the interior of patients' eyes. After seeking a couple of second opinions, the author established that what Lowell saw on Venus was actually the network of shadows cast on Lowell's own retina by his ocular blood vessels. When you compare Lowell's diagram of the spokes with a diagram of the eye, the two match up, canal for blood vessel. And when you combine the unfortunate fact that Lowell suffered from hypertension—which shows up clearly in the vessels of the eyeballs—with his will to believe, it's no surprise that he pegged Venus as well as Mars with teeming with intelligent, technologically capable inhabitants.

Alas, Lowell fared only slightly better with his search for Planet X, a planet thought to lie beyond Neptune. Planet X does not exist, as the astronomer E. Myles Standish Jr. demonstrated decisively in the mid-1990s. But Pluto, discovered at the Lowell Observatory in February 1930, some 13 years after Lowell's death, did serve as a fair approximation for a while. Within weeks of the observatory's big announcement, though, some astronomers had begun debating whether it should be classified as the ninth planet. Given our decision to display Pluto as a comet rather than as a planet in the Rose Center for Earth and Space, I've become and unwitting part of that debate myself, and I can assure you, it hasn't let up yet. Asteroid, planetoid, planetesimal, large planetesimal, icy planetesimal, minor planet, dwarf planet, giant comet, Kuiper Belt object, trans-Neptunian object, methane snowball, Mickey's dim-witted bloodhound—anything but number nine, we naysayers argue. Pluto is just too small, too lightweight, too icy, too eccentric in its orbit, too misbehaved. And by the way, we say the same about the recent high-profile contenders including the three or four objects discovered beyond Pluto that rival Pluto in size and in table manners.

TIME AND TECHNOLOGY moved on. Come the 1950s, radio-wave observations and better photography revealed fascinating facts about the planets. By the 1960s, people and robots had left Earth to take family photos of the planets. And with each new fact and photograph the curtain of ignorance lifted a bit higher.

Venus, named after the goddess of beauty and love, turns out to have a thick, almost opaque atmosphere, made up mostly of carbon dioxide, bearing down at nearly 100 times the sea level pressure on Earth. Worse yet, the surface air temperature nears 900 degrees Fahrenheit. On Venus you could cook a 16-inch pepperoni pizza in seven seconds, just by holding it out to the air. (Yes, I did the math.) Such extreme conditions pose great challenges to space exploration, because practically anything you can imagine sending to Venus will, within a moment or two, get crushed, melted, or vaporized. So you must be heatproof or just plain quick if you're collecting data from the surface of this forsaken place.

It's no accident, by the way, that Venus is hot. It suffers from a runaway greenhouse effect, induced by the carbon dioxide in its atmosphere, which traps infrared energy. So even though the tops of Venus's clouds reflect most of the Sun's incoming visible light, rocks and soils on the ground absorb the little bit that makes its way through. This same terrain then reradiates the visible light as infrared, which builds and builds in the air, eventually creating— and now sustaining—a remarkable pizza oven.

By the way, were we to find life-forms on Venus, we would probably call them Venutians, just as people from Mars would be Martians. But according to rules of Latin genitives, to be "of Venus" ought to make you a Venereal. Unfortunately, medical doctors reached that word before astronomers did. Can't blame them, I suppose. Venereal disease long predates astronomy, which itself stands as only the *second* oldest profession.

The rest of the solar system continues to become more familiar by the day. The first spacecraft to fly past Mars was *Mariner 4,* in 1965, and it sent back the first-ever close-ups of the Red Planet. Lowell's lunacies notwithstanding, before 1965 nobody knew what the Martian surface looked like, other than that it was reddish, had polar ice caps, and showed darker and lighter patches. Nobody knew it had mountains, or a canyon system vastly wider, deeper, and longer than the Grand Canyon. Nobody knew it had volcanoes vastly bigger than the largest volcano on Earth—Mauna Kea in Hawaii—even when you measure its height from the bottom of the ocean.

Nor is there any shortage of evidence that liquid water once flowed on the Martian surface: the planet has (dry) meandering riverbeds as long and wide as the Amazon, webs of (dry) tributaries, (dry) river deltas, and (dry) floodplains. The Mars exploration rovers, inching their way across the dusty rock-strewn surface, confirmed the presence of surface minerals that form only in the presence of water. Yes, signs of water everywhere, but not a drop to drink.

Something bad happened on both Mars and Venus. Could something bad happen on Earth too? Our species currently turns row upon row of environmental knobs, without much regard to long-term consequences. Who even knew to ask these questions of Earth before the study of Mars and Venus, our nearest neighbors in space, forced us to look back on ourselves?

TO GET A better view of the more distant planets requires space probes. The first spacecraft to leave the solar system were *Pioneer 10,* launched in 1972, and its twin *Pioneer 11,* launched in 1973. Both passed by Jupiter two years later, executing a grand tour along the way. They'll soon pass 10 billion miles from Earth, more than twice the distance to Pluto.

When they were launched, however, *Pioneer 10* and *11* weren't supplied with enough energy to go much beyond Jupiter. How do you get a spacecraft to go farther than its energy supply will carry it? You aim it, fire the rockets, and then just let it coast to its destination, falling along the streams of gravitational forces set up by everything in the solar system. And because astrophysicists map trajectories with precision, probes can gain energy from multiple slingshot-style maneuvers that rob orbital energy from the planets they visit. Orbital dynamicists have gotten so good at these gravity assists that they make pool sharks jealous.

Pioneer 10 and *11* sent back better pictures of Jupiter and Saturn than had ever been possible from Earth's surfce. But it was the twin spacecraft *Voyager 1* and *2*—launched in 1977 and equipped with a suite of scientific experiments and imagers—that turned the outer planets into icons. *Voyager 1* and *2* brought the solar system into the living rooms of an entire generation of world citizens. One of the windfalls of those journeys was the revelation that the moons of the outer planets are just as different from one another, and just as fascinating, as the planets themselves. Hence those planetary satellites graduated from boring points of light to worlds worthy of our attention and affection.

As I write, NASA's *Cassini* orbiter continues to orbit Saturn, in deep study of the planet itself, its striking ring system, and its many moons. Having reached Saturn's neighborhood after a "four-cushion" gravity assist, *Cassini* successfully deployed a daughter probe named *Huygens*, designed by the European Space Agency and named for Christiaan Huygens the Dutch astronomer who first identified Saturn's rings. The probe descended into the atmosphere of Saturn's largest satellite, Titan—the only moon in the solar system known to have a dense atmosphere. Titan's surface chemistry, rich in organic molecules, may be the best analog we have for the early prebiotic Earth. Other complex NASA missions are now being planned that will do the same for Jupiter, allowing a sustained study of the planet and its 70-plus moons.

IN 1584, in his book *On the Infinite Universe and Worlds,* the Italian monk and philosopher Giordano Bruno proposed the existence of "innumerable suns" and "innumerable Earths [that] revolve about these suns." Moreover, he claimed, working from the premise of a Creator both glorious and omnipotent, that each of those Earths has living inhabitants. For these and related blasphemous transgressions, the Catholic Church had Bruno burned at the stake.

Yet Bruno was neither the first nor the last person to posit some version of those ideas. His predecessors range from the fifth-century B.C. Greek philosopher Democritus to the fifteenth-century cardinal Nicholas of Cusa. His successors include such personages as the eighteenth-century philosopher Immanuel Kant and the nineteenth-century novelist Honoré de Balzac. Bruno was just unlucky to be born at a time when you could get executed for such thoughts.

During the twentieth century, astronomers figured that life could exist on other planets, as it does on Earth, only if those planets orbited their host star within the "habitable zone"—a swath of space neither too close, because water would evaporate, nor too far, because water would freeze. No doubt that life as we know it requires liquid water, but everyone had just assumed that life also required starlight as its ultimate source of energy.

Then came the discovery that Jupiter's moons Io and Europa, among other objects in the outer solar system, are heated by energy sources other than the Sun. Io is the most volcanically active place in the solar system, belching sulfurous gases into its atmosphere and spilling lava left and right. Europa almost surely has a deep billion-year-old ocean of liquid water beneath its icy crust. In both cases, the stress of Jupiter's tides on the solid moons pumps energy to their interiors, melting ice and giving rise to environments that might sustain life independent of solar energy.

Even right here on Earth, new categories of organisms, collectively called extremophiles, thrive in conditions inimical to human beings. The concept of a habitable zone incorporated an initial bias that room temperature is just right for life. But some organisms just love several-hundred-degree hot tubs and find room temperature downright hostile. To them, we are the extremophiles. Many places on Earth, previously presumed to be unlivable, such creatures call home: the bottom of Death Valley, the mouths of hot vents at the bottom of the ocean, and nuclear waste sites, to name just a few.

Armed with the knowledge that life can appear in places vastly more diverse than previously imagined, astrobiologists have broadened the earlier, and more restricted, concept of a habitable zone. Today we know that such a zone must encompass the newfound hardiness of microbial life as well as the range of energy sources that can sustain it. And, just as Bruno and others had suspected, the roster of confirmed exosolar planets continues to grow by leaps and bounds. That number has now risen past 150—all discovered in the past decade or so.

Once again we resurrect the idea that life might be everywhere, just as our ancestors had imagined. But today, we do so without risk of being immolated, and with the newfound knowledge that life is hardy and that the habitable zone may be as large as the universe itself.

EIGHT

VAGABONDS
OF THE SOLAR SYSTEM

For hundreds of years, the inventory of our celestial neigh-borhood was quite stable. It included the Sun, the stars, the planets, a handful of planetary moons, and the comets. Even the addition of a planet or two to the roster didn't change the basic organization of the system.

But on New Year's Day of 1801 a new category arose: the aster-oids, so named in 1802 by the English astronomer Sir John Herschel, son of Sir William, the discoverer of Uranus. During the next two centuries, the family album of the solar system became crammed with the data, photographs, and life histories of asteroids, as astronomers located vast numbers of these vagabonds, identified their home turf, assessed their ingredients, estimated their sizes, mapped their shapes, calculated their orbits, and crash-landed probes on them. Some investigators have also suggested that the asteroids are kinfolk to comets and even to planetary moons. And at this very moment, some astrophysicists and engineers are plot-ting methods to deflect any big ones that may be planning an unin-vited visit.

TO UNDERSTAND THE small objects in our solar system, one should look first at the large ones, specifically the planets. One curious fact about the planets is captured in a fairly simple mathematical rule proposed in 1766 by a Prussian astronomer named Johann Daniel Titius. A few years later, Titius's colleague Johann Elert Bode, giving no credit to Titius, began to spread the word about the rule, and to this day it's often called the Titius-Bode law or even, erasing Titius's contribution altogether, Bode's law. Their handy-dandy formula yielded pretty good estimates for the distances between the planets and the Sun, at least for the ones known at the time: Mercury, Venus, Earth, Mars, Jupiter, and Saturn. In 1781, widespread knowledge of the Titius-Bode law actually helped lead to the discovery of Neptune, the eighth planet from the Sun. Impressive. So either the law is just a coincidence, or it embodies some fundamental fact about how solar systems form.

It's not quite perfect, though.

Problem number 1: You have to cheat a little to get the right distance for Mercury, by inserting a zero where the formula calls for 1.5. Problem number 2: Neptune, the eighth planet, turns out to be much farther out than the formula predicts, orbiting more or less where a ninth planet should be. Problem no. 3: Pluto, which some people persist in calling the ninth planet[*] falls way off the arithmetic scale, like so much else about the place.

The law would also put a planet orbiting in the space between Mars and Jupiter—at about 2.8 astronomical units[†] from the Sun. Encouraged by the discovery of Uranus at more or less the distance Titius-Bode said it would be, astronomers in the late eighteenth century thought it would be a good idea to check out the zone around 2.8 AUs. And sure enough, on New Year's Day 1801, the Italian astronomer Giuseppe Piazzi, founder of the Observatory of Palermo, discovered something there. Subsequently it disappeared behind the glare of the Sun, but exactly one year later, with the help

---

[*]For our exhibits at the Rose Center for Earth and Space in New York City, we think of icy Pluto as one of the "kings of comets," an informative title that Pluto surely appreciates more than "puniest planet."

[†]One astronomical unit, abbreviated AU, is the average distance between Earth and the Sun.

of brilliant computations by the German mathematician Carl Friedrich Gauss, the new object was rediscovered in a different part of the sky. Everybody was excited: a triumph of mathematics and a triumph of telescopes had led to the discovery of a new planet. Piazzi himself named it Ceres (as in "cereal"), for the Roman goddess of agriculture, in keeping with the tradition of naming planets after ancient Roman deities.

But when the astronomers looked a bit harder, and calculated an orbit and a distance and a brightness for Ceres, they discovered that their new "planet" was teeny. Within a few more years three more teeny planets—Pallas, Juno, and Vesta—were discovered in the same zone. It took a few decades, but Herschel's term "asteroids" (literally "starlike" bodies) eventually caught on, because, unlike planets, which showed up in the telescopes of the day as disks, the newfound objects could not be distinguished from stars except by their motion. Further observations revealed a proliferation of asteroids, and by the end of the nineteenth century, 464 of them had been discovered in and around the swath of celestial real estate at 2.8 AU. And because the swath turned out to be a relatively flat band and did not scatter around the Sun in every direction, like bees around a hive, the zone became known as the asteroid belt.

By now, many tens of thousands of asteroids have been catalogued, with hundreds more discovered every year. Altogether, by some estimates, more than a million measure a half-mile across and up. As far as anyone can tell, even though Roman gods and goddesses did lead complicated social lives, they didn't have 10,000 friends, and so astronomers had to give up on that source of names long ago. So asteroids can now be named after actors, painters, philosophers, and playwrights; cities, countries, dinosaurs, flowers, seasons, and all manner of miscellany. Even regular people have asteroids named after them. Harriet, Jo-Ann, and Ralph each have one: they are called 1744 Harriet, 2316 Jo-Ann, and 5051 Ralph, with the number indicating the sequence in which each asteroid's orbit became firmly established. David H. Levy, a Canadian-born amateur astronomer who is the patron saint of comet hunters but has discovered plenty of asteroids as well, was kind enough to pull an asteroid from his stash and name it after me, 13123 Tyson. He did this shortly after we opened our $240-million Rose Center for Earth and Space, designed solely to bring the universe down to Earth. I was deeply moved by David's gesture, and quickly learned from 13123 Tyson's orbital data that it travels among most of the others, in the main belt of asteroids, and does not cross Earth's orbit, putting life on Earth at risk of extinction. It's just good to check this sort of thing.

ONLY CERES—the largest of the asteroids, at about 580 miles in diameter—is spherical. The others are much smaller, craggy fragments shaped like doggy bones or Idaho potatoes. Curiously, Ceres alone accounts for about a quarter of the total asteroidal mass. And if you add up the masses of all the asteroids big enough to see, plus all the smaller asteroids whose existence can be extrapolated from the data, you don't get anywhere near a planet's worth of mass. You get about 5 percent the mass of Earth's moon. So the prediction from Titius-Bode, that a red-blooded planet lurks at 2.8 AU, was a bit exaggerated.

Most asteroids are made entirely of rock, though some are entirely metal and some are a mixture of both; most inhabit what's often called the main belt, a zone between Mars and Jupiter. Asteroids are usually described as being formed of material left over from the earliest days of the solar system—material that never got incorporated into a planet. But that explanation is incomplete at best and does not account for the fact that some asteroids are pure metal. To understand what's going on, one should first consider how the larger objects in the solar system formed.

The planets coalesced from a cloud of gas and dust enriched by the scattered remains of element-rich exploding stars. The collapsing cloud forms a protoplanet—a solid blob that gets hot as it accretes more and more material. Two things happen with the larger protoplanets. One, the blob tends to take on the shape of a sphere. Two, its inner heat keeps the protoplanet molten long enough for the heavy stuff—primarily iron, with some nickel and a splash of such metals as cobalt, gold, and uranium mixed in—to sink to the center of the growing mass. Meanwhile, the much more common, light stuff—hydrogen, carbon, oxygen, and silicon— floats upward toward the surface. Geologists (who are fearless of sesquipedalian words) call the process "differentiation." Thus the core of a differentiated planet such as Earth, Mars, or Venus is metal; its mantle and crust are mostly rock, and occupy a far greater volume than the core.

Once it has cooled, if such a planet is then destroyed—say, by smashing into one of its fellow planets—the fragments of both will continue orbiting the Sun in more or less the same trajectories that the original, intact objects had. Most of those fragments will be rocky, because they come from the thick, outer, rocky layers of the two differentiated objects, and a small fraction will be purely metallic. Indeed, that's exactly what's observed with real asteroids. Moreover, a hunk of iron could not have formed in the middle of interstellar space, because the individual iron atoms of which it's made would have been scattered throughout the gas clouds that formed the planets, and gas clouds are mostly hydrogen and helium. To concentrate the iron atoms, a fluid body must first have differentiated.

BUT HOW DO solar system astronomers know that most main-belt asteroids are rocky? Or how do they know anything at all? The chief indicator is an asteroid's ability to reflect light, its albedo. Asteroids don't emit light of their own; they only absorb and reflect the Sun's rays. Does 1744 Harriet reflect or absorb infrared? What about visible light? Ultraviolet? Different materials absorb and reflect the various bands of light differently. If you're thoroughly familiar with the spectrum of sunlight (as astrophysicists are), and if you carefully observe the spectra of the sunlight reflected from an individual asteroid (as astrophysicists do), then you can figure out just how the original sunlight has been altered and thus identify the materials that comprise the asteroid's surface. And from the material, you can know how much light gets reflected. From that figure and from the distance, you can then estimate the asteroid's size. Ultimately you're trying to account for how bright an asteroid looks on the sky: it might be either really dull and big, or highly reflective and small, or something in between, and without knowing the composition, you can't know the answer simply by looking at how bright it is.

This method of spectral analysis led initially to a simplified three-way classification scheme, with carbon-rich C-type asteroids, silicate-rich S-type asteroids, and metal-rich M-type asteroids. But higher precision measurements have since spawned an alphabet soup of a dozen classes, each identifying an important nuance of the asteroid's composition and betraying multiple parent bodies rather than a single mother planet that had been smashed to smithereens.

If you know an asteroid's composition then you have some confidence that you know its density. Curiously, some measurements of the sizes of asteroids and their masses yielded densities that were less than that of rock. One logical explanation was that those asteroids weren't solid. What else could be mixed in? Ice, perhaps? Not likely. The asteroid belt sits close enough to the Sun that any species of ice (water, ammonia, carbon dioxide)—all of whose density falls below that of rock—would have evaporated long ago due to the Sun's heat. Perhaps all that's mixed in is empty space, with rocks and debris all moving in tandem.

The first bit of observational support for that hypothesis appeared in images of the 35-mile-long asteroid Ida, photographed by the space probe *Galileo* during its flyby on August 28, 1993. Half a year later a speck was spotted about 60 miles from Ida's center that proved to be a mile-wide, pebble-shaped moon! Dubbed Dactyl, it was the first satellite ever seen orbiting an asteroid. Are satellites a rare thing? If an asteroid can have a satellite orbiting it, could it have two or ten or a hundred? In other words, could some asteroids turn out to be heaps of rocks?

The answer is a resounding yes. Some astrophysicists would even say that these "rubble piles" as they are now officially named (astrophysicists once again preferred pith over polysyllabic prolixity) are probably common. One of the most extreme examples of the type may be Psyche, which measures about 150 miles in overall diameter and is reflective, suggesting its surface is metallic. From estimates of its overall density, however, its interior may well be more than 70 percent empty space.

WHEN YOU STUDY objects that live somewhere other than the main asteroid belt, you're soon tangling with the rest of the solar system's vagabonds: Earth-crossing killer asteroids, comets, and myriad planetary moons. Comets are the snowballs of the cosmos. Usually no more than a couple of miles across, they're composed of a mixture of frozen gases, frozen water, dust, and miscellaneous particles. In fact, they may simply be asteroids with a cloak of ice that never fully evaporated. The question of whether a given fragment is an asteroid or a comet might boil down to where it formed and where it's been. Before Newton published his *Principia* in 1687, in which he laid out the universal laws of gravitation, no one had any idea that comets lived and traveled among the planets, making their rounds in and out of the solar system in highly elongated orbits. Icy fragments that formed in the far reaches of the solar system, whether in the Kuiper Belt or beyond, remain shrouded in ice and, if found on a characteristic elongated path toward the Sun, will show a rarefied but highly visible trail of water vapor and other volatile gases when it swings inside the orbit of Jupiter. Eventually, after enough visits to the inner solar system (could be hundreds or even thousands) such a comet can lose all its ice, ending up as bare rock. Indeed, some, if not all, the asteroids whose orbits cross that of Earth may be "spent" comets, whose solid core remains to haunt us.

Then there are the meteorites, flying cosmic fragments that land on Earth. The fact that, like asteroids, most meteorites are made of rock and occasionally metal suggests strongly that the asteroid belt is their country of origin. To the planetary geologists who studied the growing number of known asteroids, it became clear that not all orbits hailed from the main asteroid belt.

As Hollywood loves to remind us, someday an asteroid (or comet) might collide with Earth, but that likelihood was not recognized as real until 1963, when the astrogeologist Eugene M. Shoemaker demonstrated conclusively that the vast 50,000-year-old Barringer Meteorite Crater near Winslow, Arizona, could have resulted only from a meteorite impact, and not from volcanism, or some other Earth-based geologic forces.

As we will see futher in Section 6, Shoemaker's discovery triggered a new wave of curiosity about the intersection of Earth's orbit with that of the asteroids. In the 1990s, space agencies began to track near-earth objects—comets and asteroids whose orbits, as NASA politely puts it, "allow them to enter Earth's neighborhood."

THE PLANET JUPITER plays a mighty role in the lives of the more distant asteroids and their brethren. A gravitational balancing act between Jupiter and the Sun has collected families of asteroids 60 degrees ahead of Jupiter in its solar orbit, and 60 degrees behind it, each making an equilateral triangle with Jupiter and the Sun. If you do the geometry, it places the asteroids 5.2 AU from both Jupiter and the Sun. These trapped bodies are known as the Trojan asteroids, and formally occupy what's called Lagrangian points in space. As we will see in the next chapter, these regions act like tractor beams, holding fast to asteroids that drift their way.

Jupiter also deflects plenty of comets that head toward Earth. Most comets live in the Kuiper Belt, beginning with and extending far beyond the orbit of Pluto. But any comet daring enough to pass close to Jupiter will get flung into a new direction. Were it not for

Jupiter as guardian of the moat, Earth would have been pummeled by comets far more often than it has. In fact, the Oort Cloud, which is a vast population of comets in the extreme outer solar system, named for Jan Oort, the Danish astronomer who first proposed its existence, is widely thought to be composed of Kuiper Belt comets that Jupiter flung hither and yon. Indeed, the orbits of Oort Cloud comets extend halfway to the nearest stars.

What about the planetary moons? Some look like captured asteroids, such as Phobos and Deimos, the small, dim, potato-shaped moons of Mars. But Jupiter owns several icy moons. Should those be classified as comets? And one of Pluto's moons, Charon, is not much smaller than Pluto itself. Meanwhile, both of them are icy. So perhaps they should be regarded instead as a double comet. I'm sure Pluto wouldn't mind that one either.

SPACECRAFT HAVE EXPLORED a dozen or so comets and asteroids. The first to do so was the car-sized robotic U.S. craft *NEAR Shoemaker* (NEAR is the clever acronym of Near Earth Asteroid Rendezvous), which visited the nearby asteroid Eros, not accidentally just before Valentine's Day in 2001. It touched down at just four miles an hour and, instruments intact, unexpectedly continued to send back data for two weeks after landing, enabling planetary geologists to say with some confidence that 21-mile-long Eros is an undifferentiated, consolidated object rather than a rubble pile.

Subsequent ambitious missions include *Stardust*, which flew through the coma, or dust cloud, surrounding the nucleus of a comet so that it could capture a swarm of minuscule particles in its aerogel collector grid. The goal of the mission was, quite simply, to find out what kinds of space dust are out there and to collect the particles without damaging them. To accomplish this, NASA used a wacky and wonderful substance called aerogel, the closest thing to a ghost that's ever been invented. It's a dried-out, spongelike tangle of silicon that's 99.8 percent thin air. When a particle slams in at hypersonic speeds, the particle bores its way in and gradually comes to a stop, intact. If you tried to stop the same dust grain with a catcher's mitt, or with practically anything else, the high-speed dust would slam into the surface and vaporize as it stopped abruptly. Too bad *Stardust*'s return sample crash-landed back on Earth, after a drag parachute failed to deploy.

The European Space Agency is also out there exploring comets and asteroids. The *Rosetta* spacecraft, on a 12-year mission, will explore a single comet for two years, amassing more information at close range than ever before, and will then move on to take in a couple of asteroids in the main belt.

Each of these vagabond encounters seeks to gather highly specific information that may tell us about the formation and evolution of the solar system, about the kinds of objects that populate it, about the possibility that organic molecules were transferred to Earth during impacts, or about the size, shape, and solidity of near-earth objects. And, as always, deep understanding comes not from how well you describe an object, but from how that object connects with the larger body of acquired knowledge and its moving frontier. For the solar system, that moving frontier is the search for other solar systems. What scientists want next is a thorough comparison of what we and exosolar planets and vagabonds look like. Only in this way will we know whether our home life is normal or whether we live in a dysfunctional solar family.

NINE

THE FIVE POINTS OF LAGRANGE

T he first manned spacecraft ever to leave Earth's orbit was *Apollo 8*. This achievement remains one of the most remarkable, yet unheralded firsts of the twentieth century. When that moment arrived, the astronauts fired the third and final stage of their mighty *Saturn V* rocket, rapidly thrusting the command module and its three occupants up to a speed of nearly seven miles per second. Half the energy to reach the Moon had been expended just to reach Earth's orbit.

The engines were no longer necessary after the third stage fired, except for any midcourse tuning the trajectory might require to ensure the astronauts did not miss the Moon entirely. For 90 percent of its nearly quarter-million-mile journey, the command module gradually slowed as Earth's gravity continued to tug, but ever more weakly, in the opposite direction. Meanwhile, as the astronauts neared the Moon, the Moon's force of gravity grew stronger and stronger. A spot must therefore exist, en route, where the Moon's and Earth's opposing forces of gravity balance precisely. When the command module drifted across that point in space, its speed increased once again as it accelerated toward the Moon.

If gravity were the only force to be reckoned, then this spot would be the only place in the Earth-Moon system where the opposing forces canceled each other out. But Earth and the Moon also orbit a common center of gravity, which resides about a thousand miles beneath Earth's surface, along an imaginary line connecting the centers of the Moon and Earth. When objects move in circles of any size and at any speed, they create a new force that pushes outward, away from the center of rotation. Your body feels this "centrifugal" force when you make a sharp turn in your car or when you survive amusement park attractions that turn in circles. In a classic example of these nausea-inducing rides, you stand along the edge of a large circular platter, with your back against a perimeter wall. As the contraption spins up, rotating faster and faster, you feel a stronger and stronger force pinning you against the wall. At top speeds, you can barely move against the force. That's just when they drop the floor from beneath your feet and twist the thing sideways and upside down. When I rode one of these as a kid, the force was so great that I could barely move my fingers, they being stuck to the wall along with the rest of me.

If you actually got sick on such a ride, and turned your head to the side, the vomit would fly off at a tangent. Or it might get stuck to the wall. Worse yet, if you didn't turn your head, it might not make it out of your mouth due to the extreme centrifugal forces acting in the opposite direction. (Come to think of it, I haven't seen this particular ride anywhere lately. I wonder if they've been outlawed.)

Centrifugal forces arise as the simple consequence of an object's tendency to travel in a straight line after being set in motion, and so are not true forces at all. But you can calculate with them as though they are. When you do, as did the brilliant eighteenth-century French mathematician Joseph-Louis Lagrange (1736–1813), you discover spots in the rotating Earth-Moon system where the gravity of Earth, the gravity of the Moon, and the centrifugal forces of the rotating system balance. These special locations are known as the points of Lagrange. And there are five of them.

The first point of Lagrange (affectionately called L1) falls between Earth and the Moon, slightly closer to Earth than the point of pure gravitational balance. Any object placed there can orbit the Earth-Moon center of gravity with the same monthly period as the

Moon and will appear to be locked in place along the Earth-Moon line. Although all forces cancel there, this first Lagrangian point is a precarious equilibrium. If the object drifts sideways in any direction, the combined effect of the three forces will return it to its former position. But if the object drifts directly toward or away from Earth, ever so slightly, it will irreversibly fall either toward Earth or the Moon, like a barely balanced marble atop a steep hill, a hair's-width away from rolling down one side or the other.

The second and third Lagrangian points (L2 and L3) also lie on the Earth-Moon line, but this time L2 lies far beyond the far side of the Moon, while L3 lies far beyond Earth in the opposite direction. Once again, the three forces—Earth's gravity, the Moon's gravity, and the centrifugal force of the rotating system—cancel in concert. And once again, an object placed in either spot can orbit the Earth-Moon center of gravity with the same monthly period as the Moon.

The gravitational hilltops represented by L2 and L3 are much broader than the one represented at L1. So if you find yourself drifting down to Earth or the Moon, only a tiny investment in fuel will bring you right back to where you were.

While L1, L2, and L3 are respectable space places, the award for best Lagrangian points must go to L4 and L5. One of them lives far off to the left of the Earth-Moon centerline while the other is far off to the right, each representing a vertex of an equilateral triangle, with Earth and Moon serving as the other vertices.

At L4 and L5, as with their first three siblings, all forces balance. But unlike the other Lagrangian points, which enjoy only unstable equilibrium, the equilibria at L4 and L5 are stable; no matter which direction you lean, no matter which direction you drift, the forces prevent you from leaning farther, as though you were in a valley surrounded by hills.

For each of the Lagrangian points, if your object is not located exactly where all forces cancel, then its position will oscillate around the point of balance in paths called librations. (Not to be confused with the particular spots on Earth's surface where one's mind oscillates from ingested libations.) These librations are equivalent to the back-and-forth rocking a ball would undergo after rolling down a hill and overshooting the bottom.

More than just orbital curiosities, L4 and L5 represent special places where one might build and establish space colonies. All you need do is ship raw construction materials to the area (mined not only from Earth, but perhaps from the Moon or an asteroid), leave them there with no risk of drifting away, and return later with more supplies. After all the raw materials were collected in this zero-gravity environment, you could build an enormous space station—tens of miles across—with very little stress on the construction materials. And by rotating the station, the induced centrifugal forces could simulate gravity for its hundreds (or thousands) of residents. The space enthusiasts Keith and Carolyn Henson founded the "L5 Society" in August 1975 for just that purpose, although the society is best remembered for its resonance with the ideas of Princeton physics professor and space visionary Gerard K. O'Neill, who promoted space habitation in his writings such as the 1976 classic *The High Frontier: Human Colonies in Space*. The L5 Society was founded on one guiding principle: "to disband the Society in a mass meeting at L5," presumably inside a space habitat, thereby declaring "mission accomplished." In April 1987, the L5 Society merged with the National Space Institute to become the National Space Society, which continues today.

The idea of locating a large structure at libration points appeared as early as 1961 in Arthur C. Clarke's novel *A Fall of Moondust*. Clarke was no stranger to special orbits. In 1945, he was the first to calculate, in a four-page, hand-typed memorandum, the location above Earth's surface where a satellite's period exactly matches the 24-hour rotation period of Earth. A satellite with that orbit would appear to "hover" over Earth's surface and serve as an ideal relay station for radio communications from one nation to another. Today, hundreds of communication satellites do just that.

Where is this magical place? It's not low Earth orbit. Occupants there, such as the *Hubble Space Telescope* and the *International Space*

Station, take about 90 minutes to circle Earth. Meanwhile, objects at the distance of the Moon take about a month. Logically, an intermediate distance must exist where an orbit of 24 hours can be sustained. That happens to lie 22,300 miles above Earth's surface.

ACTUALLY, THERE IS NOTHING unique about the rotating Earth-Moon system. Another set of five Lagrangian points exist for the rotating Sun-Earth system. The Sun-Earth L2 point in particular has become the darling of astrophysics satellites. The Sun-Earth Lagrangian points all orbit the Sun-Earth center of gravity once per Earth year. At a million miles from Earth, in the direction opposite that of the Sun, a telescope at L2 earns 24 hours of continuous view of the entire night sky because Earth has shrunk to insignificance. Conversely, from low Earth orbit, the location of the *Hubble* telescope, Earth is so close and so big in the sky, that it blocks nearly half the total field of view. The *Wilkinson Microwave Anisotropy Probe* (named for the late Princeton physicist David Wilkinson, a collaborator on the project) reached L2 for the Sun-Earth system in 2002, and has been busily taking data for several years on the cosmic microwave background—the omnipresent signature of the big bang itself. The hilltop for the Sun-Earth L2 region in space is even broader and flatter than that for the Earth-Moon L2. By saving only 10 percent of its total fuel, the space probe has enough to hang around this point of unstable equilibrium for nearly a century.

The *James Webb Telescope*, named for a former head of NASA from the 1960s, is now being planned by NASA as the follow-on to the *Hubble*. It too will live and work at the Sun-Earth L2 point. Even after it arrives, plenty of room will remain—tens of thousands of square miles—for more satellites to come.

Another Lagrangian-loving NASA satellite, known as *Genesis*, librates around the Sun-Earth L1 point. In this case, L1 lies a million miles toward the Sun. For two and a half years, *Genesis* faced the Sun and collected pristine solar matter, including atomic and molecular particles from the solar wind. The material was then returned to Earth via a midair recovery over Utah and studied for its composition, just like the sample return of the *Stardust* mission, which had collected comet dust. *Genesis* will provide a window to the contents of the original solar nebula from which the Sun and planets formed. After leaving L1, the returned sample did a loop-the-loop around L2 and positioned its trajectory before it returned to Earth.

Given that L4 and L5 are stable points of equilibrium, one might suppose that space junk would accumulate near them, making it quite hazardous to conduct business there. Lagrange, in fact, predicted that space debris would be found at L4 and L5 for the gravitationally powerful Sun-Jupiter system. A century later, in 1905, the first of the "Trojan" family of asteroids was discovered. We now know that for L4 and L5 of the Sun-Jupiter system, thousands of asteroids lead and follow Jupiter around the Sun, with periods that equal that of Jupiter's. Behaving for all the world as though they were responding to tractor beams, these asteroids are eternally tethered by the gravitational and centrifugal forces of the Sun-Jupiter system. Of course, we expect space junk to accumulate at L4 and L5 of the Sun-Earth system as well as the Earth-Moon system. It does. But not nearly to the extent of the Sun-Jupiter encounter.

As an important side benefit, interplanetary trajectories that begin at Lagrangian points require very little fuel to reach other Lagrangian points or even other planets. Unlike a launch from a planet's surface, where most of your fuel goes to lift you off the ground, launching from a Lagrangian point would resemble a ship leaving dry dock, gently cast adrift into the ocean with only a minimal investment of fuel. In modern times, instead of thinking about self-sustained Lagrangian colonies of people and farms, we can think of Lagrangian points as gateways to the rest of solar system. From the Sun-Earth Lagrangian points you are halfway to Mars; not in distance or time but in the all-important category of fuel consumption.

In one version of our space-faring future, imagine fuel stations at every Lagrangian point in the solar system, where travelers fill up their rocket gas tanks en route to visit friends and relatives elsewhere among the planets. This travel model, however futuristic it reads, is not entirely far-fetched. Note that without fueling stations scattered liberally across the United States, your automobile would require the proportions of the *Saturn V* rocket to drive coast to coast: most of your vehicle's size and mass would be fuel, used primarily to transport the yet-to-be-consumed fuel during your cross-country trip. We don't travel this way on Earth. Perhaps the time is overdue when we no longer travel that way through space.

TEN

ANTIMATTER MATTERS

Particle physics gets my vote as the subject with the most comical jargon in the physical sciences. Where else could a neutral vector boson be exchanged between a negative muon and a muon neutrino? Or how about the gluon that gets exchanged between a strange quark and a charmed quark? Alongside these seemingly countless particles with peculiar names is a parallel universe of *anti*particles that are collectively known as antimatter. In spite of its continued appearance in science fiction stories, antimatter is decidedly nonfiction. And yes, it does tend to annihilate on contact with ordinary matter.

The universe reveals a peculiar romance between antiparticles and particles. They can be born together out of pure energy, and they can die together (annihilate) as their combined mass gets reconverted back to energy. In 1932, the American physicist Carl David Anderson discovered the antielectron, the positively charged antimatter counterpart to the negatively charged electron. Since then, antiparticles of all varieties have been routinely made in the world's particle accelerators, but only recently have antiparticles been assembled into whole atoms. An international group led by Walter Oelert of the Institute for Nuclear Physics Research in Jülich, Germany, has created atoms where an antielectron was happily bound to an antiproton. Meet antihydrogen. These first anti-atoms were created in the particle accelerator of the European Organization for Nuclear Research (better known by its French acronym CERN) in Geneva, Switzerland, where many modern contributions to particle physics have occurred.

The method is simple: create a bunch of antielectrons and a bunch of antiprotons, bring them together at a suitable temperature and density, and hope that they combine to make atoms. In the first round of experiments, Oelert's team produced nine atoms of antihydrogen. But in a world dominated by ordinary matter, life as an antimatter atom can be precarious. The antihydrogen survived for less than 40 nanoseconds (40 billionths of a second) before annihilating with ordinary atoms.

The discovery of the antielectron was one of the great triumphs of theoretical physics, for its existence had been predicted just a few years earlier by the British-born physicist Paul A. M. Dirac. In his equation for the energy of an electron, Dirac noticed two sets of solutions: one positive and one negative. The positive solution accounted for the observed properties of the ordinary electron, but the negative solution initially defied interpretation—it had no obvious correspondence to the real world.

Equations with double solutions are not unusual. One of the simplest examples is the answer to the question, "What number times itself equals nine?" Is it 3 or -3? Of course, the answer is both, because $3 \times 3 = 9$ and $-3 \times -3 = 9$. Equations carry no guarantee that their solutions correspond to events in the real world, but if a mathematical model of a physical phenomenon is correct, then manipulating its equations can be as useful as (and much easier than) manipulating the entire universe. As in the case of Dirac and antimatter, such steps often lead to verifiable predictions, and if the predictions cannot be verified, then the theory must be discarded. Regardless of the physical outcome, a mathematical model ensures that the conclusions you might draw are logical and internally consistent.

QUANTUM THEORY, also known as quantum physics, was developed in the 1920s and is the subfield of physics that describes matter on the scale of atomic and subatomic particles. Using the newly established quantum rules, Dirac postulated that occasionally a phantom electron from the "other side" might pop into this world as an ordinary electron, thus leaving behind a hole in the sea of negative energies. The hole, Dirac suggested, would experimentally reveal itself as a positively charged antielectron, or what has come to be known as a positron.

Subatomic particles have many measurable features. If a particular property can have an opposite value, then the antiparticle version will have the opposite value but will otherwise be identical. The most obvious example is electric charge: the positron resembles the electron except that the positron has a positive charge while the electron has a negative one. Similarly, the antiproton is the oppositely charged, antiparticle of the proton.

Believe it or not, the chargeless neutron also has an antiparticle. It's called—you guessed it—the antineutron. The antineutron is endowed with an opposite zero charge to the ordinary neutron. This arithmetic magic derives from the particular triplet of fractionally charged particles (quarks) that compose neutrons. The quarks that compose the neutron have charges $-1/3$, $-1/3$, $+2/3$, while those in the antineutron have $1/3$, $1/3$, $-2/3$. Each set of three add to zero net charge yet, as you can see, the corresponding components have opposite charges.

Antimatter can seem to pop into existence out of thin air. If a pair of gamma rays have sufficiently high energy, they can interact and spontaneously transform themselves into an electron-positron pair, thus converting a lot of energy into a little bit of matter as described by the famous 1905 equation of Albert Einstein:

$$E = mc^2$$

which, in plain English reads

$$Energy = (mass) \times (speed\ of\ light)^2$$

which, in even plainer English reads

$$Energy = (mass) \times (a\ very\ big\ number)$$

In the language of Dirac's original interpretation, the gamma ray kicked an electron out of the domain of negative energies to create an ordinary electron and an electron hole. The reverse is also possible. If a particle and an antiparticle collide, they will annihilate by refilling the hole and emitting gamma rays. Gamma rays are the sort of radiation you should avoid. Want proof? Just remember how the comic strip character "The Hulk" became big, green, and ugly.

If you somehow managed to manufacture a blob of antiparticles at home, you would immediately have a storage problem, because your antiparticles would annihilate with any conventional sack or grocery bag (either paper or plastic) in which you chose to carry them. A cleverer solution traps the charged antiparticles within the confines of a strong magnetic field, where they are repelled by the magnetic walls. With the magnetic field embedded in a vacuum, the antiparticles are also rendered safe from annihilation with ordinary matter. This magnetic equivalent of a bottle is also the bag of choice when handling other container-hostile materials such as the 100-million-degree glowing gases of (controlled) nuclear fusion experiments. The real storage problem arises after you have created whole (and therefore electrically neutral) anti-atoms, because they do not normally rebound from a magnetic wall. It would be best to keep your positrons and antiprotons separate until absolutely necessary.

IT TAKES AT least as much energy to generate antimatter as you recover when it annihilates to become energy again. Unless you had a full tank of fuel in advance, self-generating antimatter engines would slowly suck energy from your starship. I don't know whether they knew about this on the original *Star Trek* television and film series but I seem to remember that Captain Kirk was always asking for "more power" from the matter-antimatter drives and Scotty was always saying that "the engines can't take it."

While there is no reason to expect a difference, the properties of antihydrogen have not yet been shown to be identical to the corresponding properties of ordinary hydrogen. Two obvious things to check are the detailed behavior of the positron in the bound company of an antiproton—does it obey all the laws of quantum theory? And the strength of an anti-atom's force of gravity—does it exhibit antigravity instead of ordinary gravity? On the atomic scales, the force of gravity between particles is unmeasurably small. Actions are instead dominated by atomic and nuclear forces, both of which are much, much stronger than gravity. What you need are enough anti-atoms to make ordinary-sized objects so that their bulk properties can be measured and compared with ordinary matter. If a set of billiard balls (and, of course, the billiard table and the cue sticks) were made of antimatter, would a game of antipool be indistinguishable from a game of pool? Would an anti-eightball fall to Earth at exactly the same rate as an ordinary eightball? Would antiplanets orbit an antistar in exactly the same way that ordinary planets orbit ordinary stars?

I am philosophically convinced that the bulk properties of antimatter will prove to be identical to those of ordinary matter—normal gravity, normal collisions, normal light, normal pool sharking, etc. Unfortunately, this means that an antigalaxy on a collision course with the Milky Way would be indistinguishable from an ordinary galaxy until it was too late to do anything about it. But this fearsome fate cannot be common in the universe because, for example, if a single antistar annihilated with a single ordinary star, then the conversion of matter to gamma-ray energy would be swift and total. Two stars with masses similar to that of the Sun (each with about 10^{57} particles) would become so luminous that the colliding system would temporarily outproduce all the energy of all the stars of a hundred million galaxies. There is no compelling evidence that such an event has ever occurred. So, as best as we can judge, the universe is dominated by ordinary matter. In other words, being annihilated need not be one of your safety concerns on that next intergalactic voyage.

Still, the universe remains disturbingly imbalanced: when created, every antiparticle is always accompanied by its particle counterpart, yet ordinary particles seem to be perfectly happy without their antiparticles. Are there hidden pockets of antimatter in the universe that account for the imbalance? Was a law of physics violated (or an unknown law of physics at work) during the early universe that forever tipped the balance in favor of matter over antimatter? We may never know the answers to these questions, but in the meantime, if an alien lands on your front lawn and extends an appendage as a gesture of greeting, before you get friendly, toss it an eightball. If the appendage explodes, then the alien was probably made of antimatter. If not, then you can proceed to take it to your leader.

WAYS
AND
MEANS
OF
NATURE

HOW NATURE PRESENTS HERSELF
TO THE INQUIRING MIND

ELEVEN

THE IMPORTANCE
OF BEING CONSTANT

Mention the word "constant," and your listeners may think of matrimonial fidelity or financial stability—or maybe they'll declare that change is the only constant in life. As it happens, the universe has its own constants, in the form of unvarying quantities that endlessly reappear in nature and in mathematics, and whose exact numerical values are of signal importance to the pursuit of science. Some of these constants are physical, grounded in actual measurements. Others, though they illuminate the workings of the universe, are purely numerical, arising from within mathematics itself.

Some constants are local and limited, applicable in just one context, one object, or one subgroup. Others are fundamental and universal, relevant to space, time, matter, and energy everywhere, thereby granting investigators the power to understand and predict the past, present, and future of the universe. Scientists know of only a few fundamental constants. The top three on most people's lists are the speed of light in a vacuum, Newton's gravitational constant, and Planck's constant, the foundation of quantum physics and the key to Heisenberg's infamous uncertainty principle. Other universal constants include the charge and mass of each of the fundamental subatomic particles.

Whenever a repeating pattern of cause and effect shows up in the universe, there's probably a constant at work. But to measure cause and effect, you must sift through what is and is not variable, and you must ensure that a simple correlation, however tempting it may be, is not mistaken for a cause. In the 1990s the stork population of Germany increased and the German at-home birth rate rose as well. Shall we credit storks for airlifting the babies? I don't think so.

But once you're certain that the constant exists, and you've measured its value, you can make predictions about places and things and phenomena yet to be discovered or thought of.

JOHANNES KEPLER, a German mathematician and occasional mystic, made the first-ever discovery of an unchanging physical quantity in the universe. In 1618, after a decade of engaging in mystical drivel, Kepler figured out that if you square the time it takes a planet to go around the Sun, then that quantity is always proportional to the cube of the planet's average distance from the Sun. Turns out, this amazing relation holds not only for each planet in our solar system but also for each star in orbit around the center of its galaxy, and for each galaxy in orbit around the center of its galactic cluster. As you might suspect, though, unbeknownst to Kepler, a constant was at work: Newton's gravitational constant lurked within Kepler's formulas, not to be revealed as such for another 70 years.

Probably the first constant you learned in school was pi—a mathematical entity denoted, since the early eighteenth century, by the Greek letter π. Pi is, quite simply, the ratio of the circumference of a circle to its diameter. In other words, pi is the multiplier if you want to go from a circle's diameter to its circumference. Pi also pops up in plenty of popular and peculiar places, including the areas of circles and ellipses, the volumes of certain solids, the motions of pendulums, the vibrations of strings, and the analysis of electrical circuits.

Not a whole number, pi instead has an unlimited succession of nonrepeating decimal digits; when truncated to include every Arabic numeral, pi looks like 3.14159265358979323846264338327950. No matter when or where you live, no matter your nationality or age or aesthetic proclivities, no matter your religion or whether you vote Democrat or Republican, if you calculate the value of pi you will get the same answer as everybody else in the universe. Constants such as pi enjoy a level of internationality that human affairs do not, never did, and never will—which is why, if people ever do communicate with aliens, they're likely to speak in mathematics, the lingua franca of the cosmos.

So we call pi an "irrational" number. You can't represent the exact value of pi as a fraction made up of two whole numbers, such as 2/3 or 18/11. But the earliest mathematicians, who had no clue about the existence of irrational numbers, didn't get much beyond representing pi as 25/8 (the Babylonians, about 2000 B.C.) or 256/81 (the Egyptians, about 1650 B.C.). Then, in about 250 B.C., the Greek mathematician Archimedes—by engaging in a laborious geometric exercise—came up with not one fraction but two, 223/71 and 22/7. Archimedes realized that the exact value of pi, a value he himself did not claim to have found, had to lie somewhere in between.

Given the progress of the day, a rather poor estimate of pi also appears in the Bible, in a passage describing the furnishings of King Solomon's temple: "a molten sea, ten cubits from the one brim to the other: it was round all about . . . and a line of thirty cubits did compass it round about" (1 Kings 7:23). That is, the diameter was 10 units, and the circumference 30, which can only be true if pi were equal to 3. Three millennia later, in 1897, the lower house of the Indiana State Legislature passed a bill announcing that, henceforth in the Hoosier state, "the ratio of the diameter and circumference is as five-fourths to four"—in other words, exactly 3.2.

Decimal-challenged lawmakers notwithstanding, the greatest mathematicians—including Muhammad ibn Musa al-Khwarizmi, a ninth-century Iraqi whose name lives on in the word "algorithm," and even Newton—steadily labored to increase the precision of pi. The advent of electronic computers, of course, blew the roof right off that exercise. As of the early twenty-first century, the number of known digits of pi has passed the 1 trillion mark, surpassing any physical application except the study (by pi-people) of whether the sequence of numerals will ever look random.

OF FAR MORE importance than Newton's contribution to the calculation of pi are his three universal laws of motion and his single universal law of gravitation. All four laws were first presented in his master work, *Philosophiæ Naturalis Principia Mathematica,* or the *Principia,* for short, published in 1687.

Before Newton's *Principia,* scientists (concerned with what was then called mechanics, and later called physics) would simply describe what they saw, and hope that the next time around it would happen the same way. But armed with Newton's laws of motion, they could describe the relations among force, mass, and acceleration under all conditions. Predictability had entered science. Predictability had entered life.

Unlike his first and third laws, Newton's second law of motion is an equation:

$$F = ma$$

Translated into English, that means a net force *(F)* applied to an object of a given mass *(m)* will result in the acceleration *(a)* of that object. In even plainer English, a big force yields a big acceleration. And they change in lockstep: double the force on an object, and you double its acceleration. The object's mass serves as the equation's constant, enabling you to calculate exactly how much acceleration you can expect from a given force.

But suppose an object's mass is not constant? Launch a rocket, and its mass drops continuously until the fuel tanks run out. And now, just for grins, suppose the mass changes even though you neither add nor subtract material from the object. That's what happens in Einstein's special theory of relativity. In the Newtonian universe, every object has a mass that is always and forever its mass. In the Einsteinian, relativistic universe, by contrast, objects have an unchanging "rest mass" (the same as the "mass" in Newton's equations), to which you add more mass according to the object's speed. What's going on is that as you accelerate an object in Einstein's universe, its resistance to that acceleration increases, showing up in the equation as an increase in the object's mass. Newton could not have known about these "relativistic" effects, because they become significant only at speeds comparable to the speed of light. To Einstein, they meant some other constant was at work: the speed of light, a subject worthy of its own essay at another time.

AS IS TRUE for many physical laws, Newton's laws of motion are plain and simple. His universal law of gravitation is somewhat more complicated. It declares that the strength of the gravitational attraction between two objects—whether between an airborne cannonball and Earth, or the Moon and Earth, or two atoms, or two galaxies—depends only on the two masses and the distance between them. More precisely, the force of gravity is directly proportional to the mass of one object times the mass of the other, and inversely proportional to the square of the distance between them. Those proportionalities give deep insight into how nature works: if the strength of the gravitational attraction between two bodies happens to be some force F at one distance, it becomes one-fourth F at double the distance and one-ninth F when the distance is tripled.

But that information by itself is not enough to calculate the exact values of the forces at work. For that, the relation requires a constant—in this case, a term known as the gravitational constant G, or, among people on the friendliest terms with the equation, "big G."

Recognizing the correspondence between distance and mass was one of Newton's many brilliant insights, but Newton had no way to measure the value of G. To do so, he would have had to know everything else in the equation, leaving G fully determined. In Newton's day, however, you could not know the whole equation. Although you could easily measure the mass of two cannonballs and their distance from each other, their mutual force of gravity would be so small that no available apparatus could have detected it. You might measure the force of gravity between Earth and a cannonball, but you had no way to measure the mass of Earth itself. Not until 1798, more than a century after the *Principia*, did the English chemist and physicist Henry Cavendish come up with a reliable measure of G.

To make his now-famous measurement, Cavendish used an apparatus whose central feature was a dumbbell, made with a pair of two-inch-diameter lead balls. A thin, vertical wire suspended the dumbbell from its middle, allowing the apparatus to twist back and forth. Cavendish enclosed the entire gizmo in an airtight case, and placed two 12-inch-diameter lead balls kitty-corner outside the case. The gravitational pull of the outside balls would tug on the dumbbell and twist the wire from which it was suspended. Cavendish's best value for G was barely accurate to four decimal places at the end of a string of zeroes. In units of cubic meters per kilogram per second squared, the value was 0.00000000006754.

Coming up with a good design for an apparatus wasn't exactly easy. Gravity is such a weak force that practically anything, even gentle air currents within the laboratory encasement, would swamp gravity's signature in the experiment. In the late nineteenth century the Hungarian physicist Loránd Eötvös, using a new and improved Cavendish-type apparatus, made mild improvements in G's precision. This experiment is so hard to do that, even today, G has acquired only a few additional decimal places. Recent experiments conducted at the University of Washington in Seattle by Jens H. Gundlach and Stephen M. Merkowitz, who redesigned the experiment, derive the value 0.000000000066742. Talk about weak: as Gundlach and Merkowitz note, the gravitational force they had to measure is equivalent to the weight of a single bacterium.

Once you know G, you can derive all kinds of things, such as Earth's mass, which had been Cavendish's ultimate goal. Gundlach and Merkowitz's best value for that is just about 5.9722 x 10^{24} kilograms, very close to the modern value.

MANY PHYSICAL CONSTANTS discovered in the past century link with forces that influence subatomic particles—a realm ruled by probability rather than precision. The most important constant among them was promulgated in 1900 by the German physicist Max Planck. Planck's constant, represented by the letter *h,* was the founding discovery of quantum mechanics, but Planck came up with it while investigating what sounds mundane: the relation between the temperature of an object and the range of energy it emits.

An object's temperature directly measures the average kinetic energy of its jiggling atoms or molecules. Of course, within this average some of the particles jiggle very fast, whereas others jiggle relatively slow. All this activity emits a sea of light, spread over a range of energies, just like the particles that emitted them. When the temperature gets high enough, the object begins to glow visibly. In Planck's day, one of the biggest challenges in physics was to explain the full spectrum of this light, particularly the bands with the highest energy.

Planck's insight was that you could account for the full sweep of the emitted spectrum in one equation only if you assume that energy itself is quantized, or divided up into itty bitty units that cannot be subdivided further: quanta.

Once Planck introduced *h* into his equation for an energy spectrum, his constant began to appear everywhere. One good place to find *h* is in the quantum description and understanding of light. The higher the frequency of light, the higher its energy: Gamma rays, the band with the highest frequencies, are maximally hostile to life. Radio waves, the band with the lowest frequencies, pass through you every second of every day, no harm done. High-frequency radiation can harm you precisely because it carries more energy. How much more? In direct proportion to the frequency.

What reveals the proportionality? Planck's constant, *h*. And if you think *G* is a minuscule constant of proportionality, take a look at the current best value for *h* (in its native kilogram-meters squared per second): 0.000000000000000000000000000000000066260693.

One of the most provocative and wondrous ways *h* appears in nature arises from the so-called uncertainty principle, first articulated in 1927 by the German physicist Werner Heisenberg. The uncertainty principle sets forth the terms of an inescapable cosmic trade-off: for various related pairs of fundamental, variable physical attributes—location and speed, energy and time—it is impossible to measure both quantities exactly. In other words, if you reduce the indeterminacy for one member of the pair (location, for instance), you're going to have to settle for a looser approximation of its partner (speed). And it's *h* that sets the limit on the precision you can attain. The trade-offs don't have much practical effect when you're measuring things in ordinary life. But when you get down to atomic dimensions, *h* rears its profound little head all around you.

IT MAY SOUND more than a bit contradictory, or even perverse, but in recent decades scientists have been looking for evidence that constants don't hold for all eternity. In 1938 the English physicist Paul A. M. Dirac proposed that the value of no less a constant than Newton's *G* might decrease in proportion to the age of the universe. Today there's practically a cottage industry of physicists desperately seeking fickle constants. Some are looking for a change across time; others, for the effects of a change in location; still others are exploring how the equations operate in previously untested domains. Sooner or later, they're going to get some real results. So stay tuned: news of inconstancy may lie ahead.

TWELVE

SPEED LIMITS

Including the space shuttle and Superman, a few things in life travel faster than a speeding bullet. But nothing moves faster than the speed of light in a vacuum. Nothing. Although as fast as light moves, its speed is decidedly not infinite. Because light has a speed, astrophysicists know that looking out in space is the same as looking back in time. And with a good estimate for the speed of light, we can come close to a reasonable estimate for the age of the universe.

These concepts are not exclusively cosmic. True, when you flick on a wall switch, you don't have to wait around for the light to reach the floor. Some morning while you're eating breakfast and you need something new to think about, though, you might want to ponder the fact that you see your kids across the table not as they are but as they once were, about three nanoseconds ago. Doesn't sound like much, but stick the kids in the nearby Andromeda galaxy, and by the time you see them spoon their Cheerios they will have aged more than 2 million years.

Minus its decimal places, the speed of light through the vacuum of space, in Americanized units, is 186,282 miles per second—a quantity that took centuries of hard work to measure with such high precision. Long before the methods and tools of science reached maturity, however, deep thinkers had thought about the nature of light: Is light a property of the perceiving eye or an emanation from an object? Is it a bundle of particles or a wave? Does it travel or simply appear? If it travels, how fast and how far?

IN THE MID-FIFTH century B.C. a forward-thinking Greek philosopher, poet, and scientist named Empedocles of Acragas wondered if light might travel at a measurable speed. But the world had to wait for Galileo, a champion of the empirical approach to the acquisition of knowledge, to illuminate the question through experiment.

He describes the steps in his book *Dialogues Concerning Two New Sciences*, published in 1638. In the dark of night, two people, each holding a lantern whose light can be rapidly covered and uncovered, stand far apart from each other, but in full view. The first person briefly flashes his lantern. The instant the second person sees the light, he flashes his own lantern. Having done the experiment just once, at a distance of less than a mile, Galileo writes:

> *I have not been able to ascertain with certainty whether the appearance of the opposite light was instantaneous or not; but if not instantaneous it is extraordinarily rapid—I should call it momentary.* (p. 43)

Fact is, Galileo's reasoning was sound, but he stood much too close to his assistant to time the passage of a light beam, particularly with the imprecise clocks of his day.

A few decades later the Danish astronomer Ole Rømer diminished the speculation by observing the orbit of Io, the innermost moon of Jupiter. Ever since January 1610, when Galileo and his brand-new telescope first caught sight of Jupiter's four brightest and largest satellites, astronomers had been tracking the Jovian moons as they circled their huge host planet. Years of observations had shown that, for Io, the average duration of one orbit—an easily timed interval from the Moon's disappearance behind Jupiter, through its re-emergence, to the beginning of its next disappearance—was just about 42.5 hours. What Rømer discovered was that when Earth was closest to Jupiter, Io disappeared about 11 minutes earlier than expected, and when Earth was farthest from Jupiter, Io disappeared about 11 minutes later.

Rømer reasoned that Io's orbital behavior was not likely to be influenced by the position of Earth relative to Jupiter, and so surely the speed of light was to blame for any unexpected variations. The 22-minute range must correspond to the time needed for light to travel across the diameter of Earth's orbit. From that assumption, Rømer derived a speed of light of about 130,000 miles a second. That's within 30 percent of the correct answer—not bad for a first-ever estimate, and a good deal more accurate than Galileo's "If not instantaneous. . . ."

James Bradley, the third Astronomer Royal of Great Britain, laid to rest nearly all remaining doubts that the speed of light was finite. In 1725 Bradley systematically observed the star Gamma Draconis and noticed a seasonal shift in the star's position on the sky. It took him three years to figure it out, but he eventually credited the shift to the combination of Earth's continuous orbital movement and the finite speed of light. Thus did Bradley discover what is known as the aberration of starlight.

Imagine an analogy: It's a rainy day, and you're sitting inside a car stuck in dense traffic. You're bored, and so (of course) you hold a big test tube out the window to collect raindrops. If there's no wind, the rain falls vertically; to collect as much water as possible, you hold the test tube in a vertical position. The raindrops enter at the top and fall straight to the bottom.

Finally the traffic clears, and your car hits the speed limit again. You know from experience that the vertically falling rain will now leave diagonal streaks on the car's side windows. To capture the raindrops efficiently, you must now tip the test tube to the angle that matches the rain streaks on the windows. The faster the car moves, the larger the angle.

In this analogy, the moving Earth is the moving car, the telescope is the test tube, and incoming starlight, because it does not move instantaneously, can be likened to the falling rain. So to catch the light of a star, you'll have to adjust the angle of the telescope—aim it at a point that's slightly different from the actual position of the star on the sky. Bradley's observation may seem a bit esoteric, but he was the first to confirm—through direct measurement rather than by inference—two major astronomical ideas: that light has a finite speed and that Earth is in orbit around the Sun. He also improved on the accuracy of light's measured speed, giving 187,000 miles per second.

BY THE LATE nineteenth century, physicists were keenly aware that light—just like sound—propagates in waves, and they presumed that if traveling sound waves need a medium (such as air) in which to vibrate, then light waves need a medium too. How else could a wave move through the vacuum of space? This mystical medium was named the "luminiferous ether," and the physicist Albert A. Michelson, working with chemist Edward W. Morley, took on the task of detecting it.

Earlier, Michelson had invented an apparatus known as an interferometer. One version of this device splits a beam of light and sends the two parts off at right angles. Each part bounces off a mirror and returns to the beam splitter, which recombines the two beams for analysis. The precision of the interferometer enables the experimenter to make extremely fine measurements of any differences in the speeds of the two light beams: the perfect device for detecting the ether. Michelson and Morley thought that if they aligned one beam with the direction of Earth's motion and made the other transverse to it, the first beam's speed would combine with Earth's motion through the ether, while the second beam's speed would remain unaffected.

Turns out, M & M got a null result. Going in two different directions made no difference to the speed of either light beam; they returned to the beam splitter at exactly the same time. Earth's motion through the ether simply had no effect on the measured speed of

light. Embarrassing. If the ether was supposed to enable the trans-
mission of light, yet it couldn't be detected, maybe the ether didn't
exist at all. Light turned out to be self-propagating: neither medium
nor magic was needed to move a beam from one position to another
in the vacuum. Thus, with a swiftness approaching the speed of light
itself, the luminiferous ether entered the graveyard of discredited scientific ideas.

And thanks to his ingenuity, Michelson also further refined the
value for the speed of light, to 186,400 miles per second.

BEGINNING IN 1905, investigations into the behavior of light got
positively spooky. That year, Einstein published his special theory
of relativity, in which he ratcheted up M & M's null result to an
audacious level. The speed of light in empty space, he declared, is a
universal constant, no matter the speed of the light-emitting source
or the speed of the person doing the measuring.

What if Einstein is right? For one thing, if you're in a spacecraft
traveling at half the speed of light and you shine a light beam
straight ahead of the spacecraft, you and I and everybody else in the
universe who measures the beam's speed will find it to be 186,282
miles per second. Not only that, even if you shine the light out the
back, top, or sides of your spacecraft, we will all continue to measure the same speed.

Odd.

Common sense says that if you fire a bullet straight ahead from
the front of a moving train, the bullet's ground speed is the speed of
the bullet *plus* the speed of the train. And if you fire the bullet
straight backward from the back of the train, the bullet's ground
speed will be its own *minus* that of the train. All that is true for bul-
lets, but not, according to Einstein, for light.

Einstein was right, of course, and the implications are staggering.
If everyone, everywhere and at all times, is to measure the same
speed for the beam from your imaginary spacecraft, a number of

things have to happen. First of all, as the speed of your spacecraft increases, the length of everything—you, your measuring devices, your spacecraft—shortens in the direction of motion, as seen by everyone else. Furthermore, your own time slows down exactly enough so that when you haul out your newly shortened yardstick, you are guaranteed to be duped into measuring the same old constant value for the speed of light. What we have here is a cosmic conspiracy of the highest order.

IMPROVED METHODS OF measuring soon added decimal place upon decimal place to the speed of light. Indeed, physicists got so good at the game that they eventually dealt themselves out of it.

Units of speed always combine units of length and time—50 miles per hour, for instance, or 800 meters per second. When Einstein began his work on special relativity, the definition of the second was coming along nicely, but definitions of the meter were completely clunky. As of 1791, the meter was defined as one ten-millionth the distance from the North Pole to the equator along the line of longitude that passes through Paris. And after earlier efforts to make this work, in 1889 the meter was redefined as the length of a prototype bar made of platinum-iridium alloy, stored at the International Bureau of Weights and Measures in Sèvres, France, and measured at the temperature at which ice melts. In 1960, the basis for defining the meter shifted again, and the exactitude increased further: 1,650,763.73 wavelengths, in a vacuum, of light emitted by the unperturbed atomic energy-level transition $2p10$ to $5d5$ of the krypton-86 isotope. Obvious, when you think about it.

Eventually it became clear to all concerned that the speed of light could be measured far more precisely than could the length of the meter. So in 1983 the General Conference on Weights and Measures decided to define—not measure, but define—the speed of light at the latest, best value: 299,792,458 meters per second. In other words, the definition of the meter was now forced into units of the speed of light, turning the meter into exactly 1/299,792,458 of the distance light travels in one second in a vacuum. And so tomorrow, anyone who measures the speed of light even more precisely than the 1983 value will be adjusting the length of the meter, not the speed of light itself.

Don't worry, though. Any refinements in the speed of light will be too small to show up in your school ruler. If you're an average European guy, you'll still be slightly less than 1.8 meters tall. And if you're an American, you'll still be getting the same bad gas mileage in your SUV.

THE SPEED OF LIGHT may be astrophysically sacred, but it's not immutable. In all transparent substances—air, water, glass, and especially diamonds—light travels more slowly than it does in a vacuum.

But the speed of light in a vacuum is a constant, and for a quantity to be truly constant it must remain unchanged, regardless of how, when, where, or why it is measured. The light-speed police take nothing for granted, though, and in the past several years they have sought evidence of change in the 13.7 billion years since the big bang. In particular, they've been measuring the so-called fine-structure constant, which is a combination of the speed of light in a vacuum and several other physical constants, including Planck's constant, pi, and the charge of an electron.

This derived constant is a measure of the small shifts in the energy levels of atoms, which affect the spectra of stars and galaxies. Since the universe is a giant time machine, in which one can see the distant past by looking at distant objects, any change in the value of the fine-structure constant with time would reveal itself in observations of the cosmos. For cogent reasons, physicists don't expect Planck's constant or the charge of an electron to vary, and pi will certainly keep its value—which leaves only the speed of light to blame if discrepancies arise.

One of the ways astrophysicists calculate the age of the universe assumes that the speed of light has always been the same, so a variation in the speed of light anywhere in the cosmos is not just of passing interest. But as of January 2006, physicists' measurements show no evidence for a change in the fine-structure constant across time or across space.

THIRTEEN

GOING BALLISTIC

In nearly all sports that use balls, the balls go ballistic at one time or another. Whether you're playing baseball, cricket, football, golf, lacrosse, soccer, tennis, or water polo, a ball gets thrown, smacked, or kicked and then briefly becomes airborne before returning to Earth.

Air resistance affects the trajectory of all these balls, but regardless of what set them in motion or where they might land, their basic paths are described by a simple equation found in Newton's *Principia,* his seminal 1687 book on motion and gravity. Several years later, Newton interpreted his discoveries for the Latin-literate lay reader in *The System of the World,* which includes a description of what would happen if you hurled stones horizontally at higher and higher speeds. Newton first notes the obvious: the stones would hit the ground farther and farther away from the release point, eventually landing beyond the horizon. He then reasons that if the speed were high enough, a stone would travel Earth's entire circumference, never hit the ground, and return to smack you in the back of the head. If you ducked at that instant, the object would continue forever in what is commonly called an orbit. You can't get more ballistic than that.

The speed needed to achieve low Earth orbit (affectionately called LEO) is a little less than 18,000 miles per hour sideways, making the round trip in about an hour and a half. Had *Sputnik 1,* the first artificial satellite, or Yury Gagarin, the first human to travel beyond Earth's atmosphere, not reached that speed after being launched, they would have come back to Earth's surface before one circumnavigation was complete.

Newton also showed that the gravity exerted by any spherical object acts as though all the object's mass were concentrated at its center. Indeed, anything tossed between two people on Earth's surface is also in orbit, except that the trajectory happens to intersect the ground. This was as true for Alan B. Shepard's 15-minute ride aboard the Mercury spacecraft *Freedom 7*, in 1961, as it is for a golf drive by Tiger Woods, a home run by Alex Rodriguez, or a ball tossed by a child: they have executed what are sensibly called suborbital trajectories. Were Earth's surface not in the way, all these objects would execute perfect, albeit elongated, orbits around Earth's center. And though the law of gravity doesn't distinguish among these trajectories, NASA does. Shepard's journey was mostly free of air resistance, because it reached an altitude where there's hardly any atmosphere. For that reason alone, the media promptly crowned him America's first space traveler.

SUBORBITAL PATHS ARE the trajectories of choice for ballistic missiles. Like a hand grenade that arcs toward its target after being hurled, a ballistic missile "flies" only under the action of gravity after being launched. These weapons of mass destruction travel hypersonically, fast enough to traverse half of Earth's circumference in 45 minutes before plunging back to the surface at thousands of miles an hour. If a ballistic missile is heavy enough, the thing can do more damage just by falling out of the sky than can the explosion of the conventional bomb it carries in its nose.

The world's first ballistic missile was the V-2 rocket, designed by a team of German scientists under the leadership of Wernher von Braun and used by the Nazis during World War II, primarily against England. As the first object to be launched above Earth's atmosphere, the bullet-shaped, large-finned V-2 (the "V" stands for *Vergeltungswaffen,* or "vengeance weapon") inspired an entire generation of spaceship illustrations. After surrendering to the Allied forces, von Braun was brought to the United States, where in 1958 he directed the launch of *Explorer 1,* the first U.S. satellite. Shortly thereafter, he was transferred to the newly created National Aeronautics and Space Administration. There he developed the *Saturn V,* the most powerful rocket ever created, making it possible to fulfill the American dream of landing on the Moon.

While hundreds of artificial satellites orbit Earth, Earth itself orbits the Sun. In his 1543 magnum opus, *De Revolutionibus,* Nicolaus Copernicus placed the Sun in the center of the universe and asserted that Earth plus the five known planets—Mercury, Venus, Mars, Jupiter, and Saturn—executed perfect circular orbits around it. Unknown to Copernicus, a circle is an extremely rare shape for an orbit and does not describe the path of any planet in our solar system. The actual shape was deduced by the German mathematician and astronomer Johannes Kepler, who published his calculations in 1609. The first of his laws of planetary motion asserts that planets orbit the Sun in ellipses. An ellipse is a flattened circle, and the degree of flatness is indicated by a numerical quantity called eccentricity, abbreviated *e.* If *e* is zero, you get a perfect circle. As *e* increases from zero to 1, your ellipse gets more and more elongated.

Of course, the greater your eccentricity, the more likely you are to cross somebody else's orbit. Comets that plunge in from the outer solar system do so on highly eccentric orbits, whereas the orbits of Earth and Venus closely resemble circles, each with very low eccentricities. The most eccentric "planet" is Pluto, and sure enough, every time it goes around the Sun, it crosses the orbit of Neptune, acting suspiciously like a comet.

THE MOST EXTREME example of an elongated orbit is the famous case of the hole dug all the way to China. Contrary to the expectations of our geographically challenged fellow citizens, China is not opposite the United States on the globe. A straight path that connects two opposite points on Earth must pass through Earth's center. What's opposite the United States? The Indian Ocean. To avoid emerging under two miles of water, we need to learn some geography and dig from Shelby, Montana, through Earth's center, to the isolated Kerguelen Islands.

Now comes the fun part. Jump in. You now accelerate continuously in a weightless, free-fall state until you reach Earth's center— where you vaporize in the fierce heat of the iron core. But let's ignore that complication. You zoom past the center, where the force of gravity is zero, and steadily decelerate until you just reach the other side, at which time you have slowed to zero. But unless a Kerguelian grabs you, you will fall back down the hole and repeat the journey indefinitely. Besides making bungee jumpers jealous, you have executed a genuine orbit, taking about an hour and a half—just like that of the space shuttle.

Some orbits are so eccentric that they never loop back around again. At an eccentricity of exactly 1, you have a parabola, and for eccentricities greater than 1, the orbit traces a hyperbola. To picture these shapes, aim a flashlight directly at a nearby wall. The emergent cone of light will form a circle of light. Now gradually angle the flashlight upward, and the circle distorts to create ellipses of higher and higher eccentricities. When your cone points straight up, the light that still falls on the nearby wall takes the exact shape of a parabola. Tip the flashlight a bit more, and you have made a hyperbola. (Now you have something different to do when you go camping.) Any object with a parabolic or hyperbolic trajectory moves so fast that it will never return. If astrophysicists ever discover a comet with such an orbit, we will know that it has emerged from the depths of interstellar space and is on a one-time tour through the inner solar system.

NEWTONIAN GRAVITY DESCRIBES the force of attraction between any two objects anywhere in the universe, no matter where they are found, what they are made of, or how large or small they may be. For example, you can use Newton's law to calculate the past and future behavior of the Earth-Moon system. But add a third object—a third source of gravity—and you severely complicate the system's motions. More generally known as the three-body problem, this ménage à trois yields richly varied trajectories whose tracking generally requires a computer.

Some clever solutions to this problem deserve attention. In one case, called the restricted three-body problem, you simplify things by assuming the third body has so little mass compared with the other two that you can ignore its presence in the equations. With this approximation, you can reliably follow the motions of all three objects in the system. And we're not cheating. Many cases like this exist in the real universe. Take the Sun, Jupiter, and one of Jupiter's itty-bitty moons. In another example drawn from the solar system, an entire family of rocks move in stable orbits around the Sun, a half-billion miles ahead of and behind Jupiter. These are the Trojan asteroids addressed in Section 2, with each one locked (as if by sci-fi tractor beams) by the gravity of Jupiter and the Sun.

Another special case of the three-body problem was discovered in recent years. Take three objects of identical mass and have them follow each other in tandem, tracing a figure eight in space. Unlike those automobile racetracks where people go to watch cars smashing into one another at the intersection of two ovals, this setup takes better care of its participants. The forces of gravity require that for all times the system "balances" at the point of intersection, and, unlike the complicated general three-body problem, all motion occurs in one plane. Alas, this special case is so odd and so rare that there is probably not a single example of it among the hundred billion stars in our galaxy, and perhaps only a few examples in the entire universe, making the figure-eight three-body orbit an astrophysically irrelevant mathematical curiosity.

Beyond one or two other well-behaved cases, the gravitational interaction of three or more objects eventually makes their trajectories go bananas. To see how this happens, one can simulate Newton's laws of motion and gravity on a computer by nudging every object according to the force of attraction between it and every other object in the calculation. Recalculate all forces and repeat. The exercise is not simply academic. The entire solar system is a many-body problem, with asteroids, moons, planets, and the Sun in a state of continuous mutual attraction. Newton worried greatly about this problem, which he could not solve with pen and paper. Fearing the entire solar system was unstable and would eventually crash its planets into the Sun or fling them into interstellar space, he postulated, as we will see in Section 9, that God might step in every now and then to set things right.

Pierre-Simon Laplace presented a solution to the many-body problem of the solar system more than a century later, in his magnum opus, *Traité de mécanique céleste*. But to do so, he had to develop a new form of mathematics known as perturbation theory. The analysis begins by assuming that there is only one major source of gravity and that all the other forces are minor, though persistent—exactly the situation in our solar system. Laplace then demonstrated analytically that the solar system is indeed stable, and that you don't need new laws of physics to show it.

Or is it? As we will see further in Section 6, modern analysis demonstrates that on timescales of hundreds of millions of years—periods much longer than the ones considered by Laplace—planetary orbits are chaotic. A situation that leaves Mercury vulnerable to falling into the Sun, and Pluto vulnerable to getting flung out of the solar system altogether. Worse yet, the solar system might have been born with dozens of other planets, most of them now long lost to interstellar space. And it all started with Copernicus's simple circles.

WHENEVER YOU GO ballistic, you are in free fall. All of Newton's stones were in free fall toward Earth. The one that achieved orbit was also in free fall toward Earth, but our planet's surface curved out from under it at exactly the same rate as it fell—a consequence of the stone's extraordinary sideways motion. The *International Space Station* is also in free fall toward Earth. So is the Moon. And, like Newton's stones, they all maintain a prodigious sideways motion that prevents them from crashing to the ground. For those objects, as well as for the space shuttle, the wayward wrenches of spacewalking astronauts, and other hardware in LEO, one trip around the planet takes about 90 minutes.

The higher you go, however, the longer the orbital period. As noted earlier, 22,300 miles up, the orbital period is the same as Earth's rotation rate. Satellites launched to that location are geo-stationary; they "hover" over a single spot on our planet, enabling rapid, sustained communication between continents. Much higher still, at an altitude of 240,000 miles, is the Moon, which takes 27.3 days to complete its orbit.

A fascinating feature of free fall is the persistent state of weight-lessness aboard any craft with such a trajectory. In free fall you and everything around you fall at exactly the same rate. A scale placed between your feet and the floor would also be in free fall. Because nothing is squeezing the scale, it would read zero. For this reason, and no other, astronauts are weightless in space.

But the moment the spacecraft speeds up or begins to rotate or undergoes resistance from Earth's atmosphere, the free-fall state ends and the astronauts weigh something again. Every science-fiction fan knows that if you rotate your spacecraft at just the right speed, or accelerate your spaceship at the same rate as an object falls to Earth, you will weigh exactly what you do on your doctor's scale. So if your aerospace engineers felt so compelled, they could design your spaceship to simulate Earth gravity during those long, boring space journeys.

Another clever application of Newton's orbital mechanics is the slingshot effect. Space agencies often launch probes from Earth that have too little energy to reach their planetary destinations. Instead, the orbit engineers aim the probes along cunning trajectories that swing near a hefty, moving source of gravity, such as Jupiter. By falling toward Jupiter in the same direction as Jupiter moves, a probe can steal some Jovial energy during its flyby and then sling forward like a jai alai ball. If the planetary alignments are right, the probe can perform the same trick as it swings by Saturn, Uranus, or Neptune, stealing more energy with each close encounter. These are not small boosts; these are big boosts. A one-time shot at Jupiter can double a probe's speed through the solar system.

The fastest-moving stars of the galaxy, the ones that give colloquial meaning to "going ballistic," are the stars that fly past the supermassive black hole in the center of the Milky Way. A descent toward this black hole (or any black hole) can accelerate a star up to speeds approaching that of light. No other object has the power to do this. If a star's trajectory swings slightly to the side of the hole, executing a near miss, it will avoid getting eaten, but its speed will dramatically increase. Now imagine a few hundred or a few thousand stars engaged in this frenetic activity. Astrophysicists view such stellar gymnastics—detectable in most galaxy centers—as conclusive evidence for the existence of black holes.

The farthest object visible to the unaided eye is the beautiful Andromeda galaxy, which is the closest spiral galaxy to us. That's the good news. The bad news is that all available data suggest that the two of us are on a collision course. As we plunge ever deeper into each other's gravitational embrace, we will become a twisted wreck of strewn stars and colliding gas clouds. Just wait about 6 or 7 billion years.

In any case, you could probably sell seats to watch the encounter between Andromeda's supermassive black hole and ours, as whole galaxies go ballistic.

FOURTEEN

ON BEING DENSE

W hen I was in the 5th grade, a mischievous classmate asked me the question, "Which weighs more,
a ton of feathers or a ton of lead?" No, I was not
fooled, but little did I know how useful a critical
understanding of density would be to life and the universe. A common way to compute density is, of course, to take the ratio of an
object's mass to its volume. But other types of densities exist, such
as the resistance of somebody's brain to the imparting of common
sense or the number of people per square mile who live on an
exotic island such as Manhattan.

The range of measured densities within our universe is stagger-
ingly large. We find the highest densities within pulsars, where
neutrons are so tightly packed that one thimbleful would weigh
about as much as a herd of 50 million elephants. And when a rab-
bit disappears into "thin air" at a magic show, nobody tells you the
thin air already contains over 10,000,000,000,000,000,000,000,000
(ten septillion) atoms per cubic meter. The best laboratory vacuum chambers can pump down to as few as 10,000,000,000 (ten
billion) atoms per cubic meter. Interplanetary space gets down to
about 10,000,000 (ten million) atoms per cubic meter, while interstellar space is as low as 500,000 atoms per cubic meter. The award
for nothingness, however, must be given to the space between
galaxies, where it is difficult to find more than a few atoms for
every 10 cubic meters.

The range of densities in the universe spans forty-four powers of 10. If one were to classify cosmic objects by density alone, salient features would reveal themselves with remarkable clarity. For example, dense compact objects such as black holes, pulsars, and white dwarf stars all have a high force of gravity at their surfaces and readily accrete matter into a funneling disk. Another example comes from the properties of interstellar gas. Everywhere where we look in the Milky Way, and in other galaxies, gas clouds with the greatest density are sites of freshly minted stars. Our detailed understanding of the star formation process remains incomplete, but understandably, nearly all theories of star formation include explicit reference to the changing gas density as clouds collapse to form stars.

OFTEN IN ASTROPHYSICS, especially in the planetary sciences, one can infer the gross composition of an asteroid or a moon simply by knowing its density. How? Many common ingredients in the solar system have densities that are quite distinct from one another. Using the density of liquid water as a measuring unit, frozen water, ammonia, methane, and carbon dioxide (common ingredients in comets) all have a density of less than 1; rocky materials, which are common among the inner planets and asteroids, have densities between 2 and 5; iron, nickel, and several other metals that are common in the cores of planets, and also in asteroids, have densities above 8. Objects with average densities intermediate to these broad groups are normally interpreted as having a mixture of these common ingredients. For Earth we can do a little better: the speed of postearthquake sound waves through Earth's interior directly relates to the run of Earth's density from its center to the surface. The best available seismic data give a core density of around 12, dropping to an outer crustal density of around 3. When averaged together, the density of the entire Earth is about 5.5.

Density, mass, and volume (size) come together in the equation for density, so if you measure or infer any two of the quantities then you can compute the third. The planet around the sunlike, naked-eye star 51 Pegasus had its mass and orbit computed directly from the data. A subsequent assumption about whether the planet is gaseous (likely) or rocky (unlikely) allows a basic estimate of the planet's size.

Often when people claim one substance to be heavier than another, the implicit comparison is one of density, not weight. For example, the simple yet technically ambiguous statement "lead weighs more than feathers" would be understood by nearly every-body to be really a question of density. But this implicit under-standing fails in some notable cases. Heavy cream is lighter (less dense) than skim milk, and all seagoing vessels, including the 150,000-ton *Queen Mary 2*, are lighter (less dense) than water. If these statements were false, then cream and ocean liners would sink to the bottom of the liquids upon which they float.

OTHER DENSITY TIDBITS:

Under the influence of gravity, hot air does not rise simply because it's hot, but because it's less dense than the surrounding air. One could similarly declare that cool, denser air sinks, both of which must happen to enable convection in the universe.

Solid water (commonly known as ice) is less dense than liquid water. If the reverse were true, then in the winter, large lakes and rivers would freeze completely, from the bottom to the top, killing all fish. What protects the fish is the floating, less dense, upper layer of ice, which insulates the warmer waters below from the cold win-ter airs.

On the subject of dead fish, when found belly-up in your fish tank, they are, of course, temporarily less dense than their live counterparts.

Unlike any other known planet, the average density of Saturn is less than that of water. In other words, a scoop of Saturn would float in your bathtub. Knowing this, I have always wanted for my bathtub entertainment a rubber Saturn instead of a rubber ducky.

If you feed a black hole, its event horizon (that boundary beyond which light cannot escape) grows in direct proportion to its mass, which means that as a black hole's mass increases, the average density within its event horizon actually decreases. Meanwhile, as far as we can tell from our equations, the material content of a black hole has collapsed to a single point of near-infinite density at its center.

And behold the greatest mystery of them all: an unopened can of diet Pepsi floats in water while an unopened can of regular Pepsi sinks.

IF YOU WERE to double the number of marbles in a box, their density would, of course, remain the same because both the mass and the volume would double, which in combination has no net effect on the density. But objects exist in the universe whose density relative to mass and volume yields unfamiliar results. If your box contained soft, fluffy down, and you doubled the number of feathers, then ones on the bottom would become flattened. You would have doubled the mass but not the volume, and you would be left with a net increase in density. All squishable things under the influence of their own weight will behave this way. Earth's atmosphere is no exception: we find half of all its molecules packed into the lowest three miles above Earth's surface. To astrophysicists, Earth's atmosphere forms a bad influence on the quality of data, which is why you often hear about us escaping to mountaintops to conduct research, leaving as much of Earth's atmosphere below us as possible.

Earth's atmosphere ends where it blends indistinguishably with the very low density gas of interplanetary space. Normally, this blend lies several thousand miles above Earth's surface. Note that the space shuttle, the *Hubble* telescope, and other satellites that orbit within only a few hundred miles of Earth's surface would eventually fall out of orbit from the residual atmospheric air resistance if they did not receive periodic boosts. During peak solar activity, however (every 11 years) Earth's upper atmosphere receives a higher dose of solar radiation, forcing it to heat and expand. During this period the atmosphere can extend an extra thousand miles into space, thus decaying satellite orbits faster than usual.

BEFORE LABORATORY VACUUMS, air was the closest thing to nothing that anyone could imagine. Along with earth, fire, and water, air was one of the original four Aristotelian elements that composed the known world. Actually, there was a fifth element known as the "quint"-essence. Otherworldly, yet lighter than air and more ethereal than fire, the rarefied quintessence was presumed to comprise the heavens. How quaint.

We needn't look as far as the heavens to find rarefied environments. Our upper atmosphere will suffice. Beginning at sea level, air weighs about 15 pounds per square inch. So if you cookie cut a square inch of atmosphere from thousands of miles up all the way down to sea level and you put it on a scale, it would weigh 15 pounds. For comparison, a square-inch column of water requires a mere 33 feet to weigh 15 pounds. On mountaintops and high up in airplanes, the cookie-cut column of air above you is shorter and therefore weighs less. At the 14,000-foot summit of Mauna Kea, Hawaii, home to some of the world's most powerful telescopes, the atmospheric pressure drops to about 10 pounds per square inch. While observing on site, astrophysicists will intermittently breathe from oxygen tanks to retain their intellectual acuity.

Above 100 miles, where there are no known astrophysicists, the air is so rarefied that gas molecules move for a relatively long time before colliding with one another. If, between collisions, the molecules are slammed by an incoming particle, they become temporarily excited and then emit a unique spectrum of colors before their next collision. When the incoming particles are the constituents of the solar wind, such as protons and electrons, the emissions are curtains of undulating light that we commonly call aurora. When the spectrum of auroral light was first measured, it had no counterpart in the laboratory. The identity of the glowing molecules remained unknown until we learned that excited, but otherwise ordinary, molecules of nitrogen and oxygen were to blame. At sea level, their rapid collisions with each other absorb this excess energy long before they have had a chance to emit their own light.

Earth's upper atmosphere is not alone in producing mysterious lights. Spectral features in the Sun's corona long puzzled astrophysicists. An extremely rarefied place, the corona is that beautiful, fiery-looking outer region of the Sun that's rendered visible during a total solar eclipse. The new feature was assigned to an unknown element dubbed "coronium." Not until we learned that the solar corona is heated to millions of degrees did we figure out that the mystery element was highly ionized iron, a previously unfamiliar state where most of its outer electrons are stripped away and floating free in the gas.

The term "rarefied" is normally reserved for gases, but I will take the liberty to apply it to the solar system's famed asteroid belt. From movies and other descriptions, you would think it was a hazardous place, wrought with the constant threat of head-on collisions with house-sized boulders. The actual recipe for the asteroid belt? Take a mere 2.5 percent of the Moon's mass (itself, just 1/81 the mass of Earth), crush it into thousands of assorted pieces, but make sure that three-quarters of the mass is contained in just four asteroids. Then spread them all across a 100-million-mile-wide belt that tracks along a 1.5-billion-mile path around the Sun.

COMET TAILS, as tenuous and rarefied as they are, represent an increase in density by a factor of 1,000 over the ambient conditions in interplanetary space. By reflecting sunlight and re-emitting energy absorbed from the Sun, a comet tail possesses remarkable visibility given its nothingness. Fred Whipple, of the Harvard-Smithsonian Center for Astrophysics, is generally considered to be a parent of our modern understanding of comets. He has succinctly described a comet's tail as the most that has ever been made of the least. Indeed, if the entire volume of a 50-million-mile-long comet tail were compressed to the density of ordinary air, all the tail's gas would fill a half-mile cube. When the astronomically common yet deadly gas cyanogen (CN) was first discovered in comets, and when it was later announced that Earth would pass through the tail of Halley's comet during its 1910 visit to the inner solar system, gullible people were sold anticomet pills by pharmaceutical charlatans.

The core of the sun, where all its thermonuclear energy is generated, is not a place to find low-density material. But the core comprises a mere 1 percent of the Sun's volume. The average density of the entire Sun is only one-fourth that of Earth, and only 40 percent higher than ordinary water. In other words, a spoonful of Sun would sink in your bathtub, but it wouldn't sink fast. Yet in 5 billion years the Sun's core will have fused nearly all its hydrogen into helium and will shortly thereafter begin to fuse helium into carbon. Meanwhile, the luminosity of the Sun will increase a thousandfold while its surface temperature drops to half of what it is today. We know from the laws of physics that the only way an object can increase its luminosity while simultaneously getting cooler is for it to get bigger. As will be detailed in Section 5, the Sun will ultimately expand to a bulbous ball of rarefied gas that will completely fill and extend beyond the volume of Earth's orbit, while the Sun's average density falls to less than one ten-billionth of its current value. Of course Earth's oceans and atmosphere will have evaporated into space and all life will have vaporized, but that needn't concern us here. The Sun's outer atmosphere, rarefied though it will be, would nonetheless impede the motion of Earth in its orbit and force us on a relentless spiral inward toward thermonuclear oblivion.

BEYOND OUR SOLAR SYSTEM we venture into interstellar space. Humans have sent four spacecraft with enough speed to journey there: *Pioneer 10* and *11*, and *Voyager 1* and *2*. The fastest among them, *Voyager 2*, will reach the distance of the nearest star to the Sun in about 25,000 years.

Yes, interstellar space is empty. But like the remarkable visibility of rarefied comet tails in interplanetary space, gas clouds out there, with a hundred to a thousand times the ambient density, can readily reveal themselves in the presence of nearby luminous stars. Once again, when the light from these colorful nebulosities was first analyzed their spectra revealed unfamiliar patterns. The hypothetical element "nebulium" was proposed as a placeholder for our ignorance. In the late 1800s, there was clearly no spot on the periodic table of elements that could possibly be identified with nebulium. As laboratory vacuum techniques improved, and as unfamiliar spectral features became routinely identified with familiar elements, suspicions grew—and were later confirmed—that nebulium was ordinary oxygen in an extraordinary state. What state was that? The atoms were each stripped of two electrons and they lived in the near-perfect vacuum of interstellar space.

When you leave the galaxy, you leave behind nearly all gas and dust and stars and planets and debris. You enter an unimaginable cosmic void. Let's talk empty: A cube of intergalactic space, 200,000 kilometers on a side, contains about the same number of atoms as the air that fills the usable volume of your refrigerator. Out there, the cosmos not only loves a vacuum, it's carved from it.

Alas, an absolute, perfect vacuum may be impossible to attain or find. As we saw in Section 2, one of the many bizarre predictions of quantum mechanics holds that the real vacuum of space contains a sea of "virtual" particles that continually pop in and out of existence along with their antimatter counterparts. Their virtuality comes from having lifetimes that are so short that their direct existence cannot ever be measured. More commonly known as the "vacuum energy," it can act as antigravity pressure that will ultimately trigger the universe to expand exponentially faster and faster—making intergalactic space all the more rarefied.

What lies beyond?

Among those who dabble in metaphysics, some hypothesize that outside the universe, where there is no space, there is no nothing. We might call this hypothetical, zero-density place, nothing-nothing, except that we are certain to find multitudes of unre-trieved rabbits.

FIFTEEN

OVER THE RAINBOW

Whenever cartoonists draw biologists, chemists, or engineers, the characters typically wear protective white lab coats that have assorted pens and pencils poking out of the breast pocket. Astrophysicists use plenty of pens and pencils, but we never wear lab coats unless we are building something to launch into space. Our primary laboratory is the cosmos, and unless you have bad luck and get hit by a meteorite, you are not at risk of getting your clothes singed or otherwise sullied by caustic liquids spilling from the sky. Therein lies the challenge. How do you study something that cannot possibly get your clothes dirty? How do astrophysicists know anything about either the universe or its contents if all the objects to be studied are light-years away?

Fortunately, the light emanating from a star reveals much more to us than its position in the sky or how bright it is. The atoms of objects that glow lead busy lives. Their little electrons continually absorb and emit light. And if the environment is hot enough, energetic collisions between atoms can jar loose some or all of their electrons, allowing them to scatter light to and fro. All told, atoms leave their fingerprint on the light being studied, which uniquely implicates which chemical elements or molecules are responsible.

As early as 1666, Isaac Newton passed white light through a prism to produce the now-familiar spectrum of seven colors: red, orange, yellow, green, blue, indigo, and violet, which he personally named. (Feel free to call them Roy G. Biv.) Others had played with prisms before. What Newton did next, however, had no precedent. He passed the emergent spectrum of colors back through a second prism and recovered the pure white he started with, demonstrating a remarkable property of light that has no counterpart on the artist's palette; these same colors of paint, when mixed, would leave you with a color resembling that of sludge. Newton also tried to disperse the colors themselves but found them to be pure. And in spite of the seven names spectral colors change smoothly and continuously from one to the next. The human eye has no capacity to do what prisms do—another window to the universe lay undiscovered before us.

A CAREFUL INSPECTION of the Sun's spectrum, using precision optics and techniques unavailable in Newton's day, reveals not only Roy G. Biv, but narrow segments within the spectrum where the colors are absent. These "lines" through the light were discovered in 1802 by the English medical chemist William Hyde Wollaston, who naively (though sensibly) suggested that they were naturally occurring boundaries between the colors. A more complete discussion and interpretation followed with the efforts of the German physicist and optician Joseph von Fraunhofer (1787–1826), who devoted his professional career to the quantitative analysis of spectra and to the construction of optical devices that generate them. Fraunhofer is often referred to as the father of modern spectroscopy, but I might further make the claim that he was the father of astrophysics. Between 1814 and 1817, he passed the light of certain flames through a prism and discovered that the pattern of lines resembled what he found in the Sun's spectrum, which further resembled lines found in the spectra of many stars, including Capella, one of the brightest in the nighttime sky.

By the mid-1800s the chemists Gustav Kirchhoff and Robert Bunsen (of Bunsen-burner fame from your chemistry class) were

making a cottage industry of passing the light of burning substances through a prism. They mapped the patterns made by
known elements and discovered a host of new elements, including
rubidium and caesium. Each element left its own pattern of lines—
its own calling card—in the spectrum being studied. So fertile was
this enterprise that the second most abundant element in the universe, helium, was discovered in the spectrum of the Sun *before* it
was discovered on Earth. The element's name bears this history
with its prefix derived from *Helios*, "the Sun."

A DETAILED AND accurate explanation of how atoms and their
electrons form spectral lines would not emerge until the era of
quantum physics a half-century later, but the conceptual leap had
already been made: Just as Newton's equations of gravity connected
the realm of laboratory physics to the solar system, Fraunhofer connected the realm of laboratory chemistry to the cosmos. The stage
was set to identify, for the first time, what chemical elements filled
the universe, and under what conditions of temperature and pressure their patterns revealed themselves to the spectroscopist.

Among the more bone-headed statements made by armchair
philosophers, we find the following 1835 proclamation in *Cours de
la Philosophie Positive* by Auguste Comte (1798–1857):

> On the subject of stars, all investigations which are not ultimately
> reducible to simple visual observations are . . . necessarily denied to
> us. . . . We shall never be able by any means to study their chemical
> composition. . . . I regard any notion concerning the true mean tem-
> perature of the various stars as forever denied to us. (p. 16, author's
> trans.)

Quotes like that can make you afraid to say anything in print.

Just seven years later, in 1842, the Austrian physicist Christian
Doppler proposed what became known as the Doppler effect,

which is the change in frequency of a wave being emitted by an object in motion. One can think of the moving object as stretching the waves behind it (reducing their frequency) and compressing the waves in front of it (increasing their frequency). The faster the object moves, the more the light is both compressed in front of it and stretched behind it. This simple relationship between speed and frequency has profound implications. If you know what frequency was emitted, but you measure it to have a different value, the difference between the two is a direct indication of the object's speed toward or away from you. In an 1842 paper, Doppler makes the prescient statement:

> It is almost to be accepted with certainty that this [Doppler effect] will in the not too distant future offer astronomers a welcome means to determine the movements . . . of such stars which . . . until this moment hardly presented the hope of such measurements and determinations. (Schwippell 1992, pp. 46–54)

The idea works for sound waves, for light waves, and in fact, waves of any origin. (I'd bet Doppler would be surprised to learn that his discovery would one day be used in microwave-based "radar guns" wielded by police officers to extract money from people who drive automobiles above a speed limit set by law.) By 1845, Doppler was conducting experiments with musicians playing tunes on flatbed railway trains, while people with perfect pitch wrote down the changing notes they heard as the train approached and then receded.

DURING THE LATE 1800S, with the widespread use of spectrographs in astronomy, coupled with the new science of photography, the field of astronomy was reborn as the discipline of astrophysics. One of the pre-eminent research publications in my field, the *Astrophysical Journal*, was founded in 1895, and, until 1962, bore the subtitle: *An International Review of Spectroscopy and Astronomical Physics*. Even today, nearly every paper reporting observations of the universe gives either an analysis of spectra or is heavily influenced by spectroscopic data obtained by others.

To generate a spectrum of an object requires much more light than to take a snapshot, so the biggest telescopes in the world, such as the 10-meter Keck telescopes in Hawaii, are tasked primarily with getting spectra. In short, were it not for our ability to analyze spectra, we would know next to nothing about what goes on in the universe.

Astrophysics educators face a pedagogical challenge of the highest rank. Astrophysics researchers deduce nearly all knowledge about the structure, formation, and evolution of things in the universe from the study of spectra. But the analysis of spectra is removed by several levels of inference from the things being studied. Analogies and metaphors help, by linking a complex, somewhat abstract idea to a simpler, more tangible one. The biologist might describe the shape of the DNA molecule as two coils, connected to each other the way rungs on a ladder connect its sides. I can picture a coil. I can picture two coils. I can picture rungs on a ladder. I can therefore picture the molecule's shape. Each part of the description sits only one level of inference removed from the molecule itself. And they come together nicely to make a tangible image in the mind. No matter how easy or hard the subject may be, one can now talk about the science of the molecule.

But to explain how we know the speed of a receding star requires five nested levels of abstraction:

Level 0: Star
Level 1: Picture of a star
Level 2: Light from the picture of a star
Level 3: Spectrum from the light from the picture of a star
Level 4: Patterns of lines lacing the spectrum from the light from the picture of a star
Level 5: Shifts in the patterns of lines in the spectrum from the light from the picture of the star

Going from level 0 to level 1 is a trivial step that we take every time we snap a photo with a camera. But by the time your explanation reaches level 5, the audience is either befuddled or just fast sleep. That is why the public hardly ever hears about the role of spectra in cosmic discovery—it's just too far removed from the objects themselves to explain efficiently or with ease.

In the design of exhibits for a natural history museum, or for any museum where real things matter, what you typically seek are artifacts for display cases—rocks, bones, tools, fossils, memorabilia, and so forth. All these are "level 0" specimens and require little or no cognitive investment before you give the explanation of what the object is. For astrophysics displays, however, any attempt to place stars or quasars on display would vaporize the museum.

Most astrophysics exhibits are therefore conceived in level 1, leading principally to displays of pictures, some quite striking and beautiful. The most famous telescope in modern times, the *Hubble Space Telescope*, is known to the public primarily through the beautiful, full-color, high-resolution images it has acquired of objects in the universe. The problem here is that after you view such exhibits, you leave waxing poetic about the beauty of the universe yet you are no closer than before to understanding how it all works. To really know the universe requires forays into levels 3, 4, and 5. While much good science has come from the *Hubble* telescope, you would never know it from media accounts that the foundation of our cosmic knowledge continues to flow primarily from the analysis of spectra and not from looking at pretty pictures. I want people to be struck, not only from exposure to levels 0 and 1, but also from exposure to level 5, which admittedly requires a greater intellectual investment on the part of the student, but also (and perhaps especially) on the part of the educator.

IT'S ONE THING to see a beautiful color picture, taken in visible light, of a nebula in our own Milky Way galaxy. But it's another thing to know from its radio-wave spectrum that it also harbors newly formed stars of very high mass within its cloud layers. This gas cloud is a stellar nursery, regenerating the light of the universe.

It's one thing to know that every now and again, high-mass stars explode. Photographs can show you this. But x-ray and visible-light spectra of these dying stars reveal a cache of heavy elements that enrich the galaxy and are directly traceable to the constituent elements of life on Earth. Not only do we live among the stars, the stars live within us.

It's one thing to look at a poster of a pretty spiral galaxy. But it's another thing to know from Doppler shifts in its spectral features that the galaxy is rotating at 200 kilometers per second, from which we infer the presence of 100 billion stars using Newton's laws of gravity. And by the way, the galaxy is receding from us at one-tenth the speed of light as part of the expansion of the universe.

It's one thing to look at nearby stars that resemble the Sun in luminosity and temperature. But it's another thing to use hypersensitive Doppler measurements of the star's motion to infer the existence of planets in orbit around them. At press time, our catalog is rising through 200 such planets outside the familiar ones in our own solar system.

It's one thing to observe the light from a quasar at the edge of the universe. But its another thing entirely to analyze the quasar's spectrum and deduce the structure of the invisible universe, laid along the quasar's path of light as gas clouds and other obstructions take their bite our of the quasar spectra.

Fortunately, for all the magnetohydrodynamicists among us, atomic structure changes slightly under the influence of a magnetic field. This change manifests itself in the slightly altered spectral pattern caused by these magnetically afflicted atoms.

And armed with Einstein's relativistic version of the Doppler formula, we deduce the expansion rate of the entire universe from the spectra of countless galaxies near and far, and thus deduce the age and fate of the universe.

One could make a compelling argument that we know more

about the universe than the marine biologist knows about the bottom of the ocean or the geologist knows about the center of Earth. Far from an existence as powerless stargazers, modern astrophysicists are armed to the teeth with the tools and techniques of spectroscopy, enabling us all to stay firmly planted on Earth, yet finally touch the stars (without burning our fingers) and claim to know them as never before.

SIXTEEN

COSMIC WINDOWS

As noted in Section 1, the human eye is often advertised to be among the most impressive of the body's organs. Its ability to focus near and far, to adjust to a broad range of light levels, and to distinguish colors are at the top of most peoples' list of eye-opening features. But when you take note of the many bands of light that are invisible to us, then you would be forced to declare humans to be practically blind. How impressive is our hearing? Bats would clearly fly circles around us with a sensitivity to pitch that extends beyond our own by an order of magnitude. And if the human sense of smell were as good as that of dogs, then Fred rather than Fido might be the one who sniffs out contraband from airport customs searches.

The history of human discovery is characterized by the boundless desire to extend the senses beyond our inborn limits. It is through this desire that we open new windows to the universe. For example, beginning in the 1960s with the early Soviet and NASA probes to the Moon and planets, computer-controlled space probes, which we can rightly call robots, became (and still are) the standard tool for space exploration. Robots in space have several clear advantages over astronauts: they are cheaper to launch; they can be designed to perform experiments of very high precision without the interference of a cumbersome pressure suit; and they are not alive in any traditional sense of the word, so they cannot be killed in a space accident. But until computers can simulate human curiosity and human sparks of insight, and until computers can synthesize information and recognize a serendipitous discovery when it stares them in the face (and perhaps even when it doesn't), robots will remain tools designed to discover what we already expect to find.

Unfortunately, profound questions about nature can lurk among those that have yet to be asked.

The most significant improvement of our feeble senses is the extension of our sight into the invisible bands of what is collectively known as the electromagnetic spectrum. In the late nineteenth century the German physicist Heinrich Hertz performed experiments that helped to unify conceptually what were previously considered to be unrelated forms of radiation. Radio waves, infrared, visible light, and ultraviolet were all revealed to be cousins in a family of light that simply differed in energy. The full spectrum, including all parts discovered after Hertz's work, extends from the low-energy part that we call radio waves, and continues in order of increasing energy to microwaves, infrared, visible (comprising the "rainbow seven": red, orange, yellow, green, blue, indigo, and violet), ultraviolet, x-rays, and gamma rays.

Superman, with his x-ray vision, has no special advantage over modern-day scientists. Yes, he is a bit stronger than your average astrophysicist, but astrophysicists can now "see" into every major part of the electromagnetic spectrum. In the absence of this extended vision we are not only blind but ignorant—the existence of many astrophysical phenomena reveal themselves only through some windows and not others.

WHAT FOLLOWS IS a selective peek through each window to the universe, beginning with radio waves, which require very different detectors from those you will find in the human retina.

In 1932 Karl Jansky, in the employ of Bell Telephone Laboratories and armed with a radio antenna, first "saw" radio signals that emanated from somewhere other than Earth; he had discovered the center of the Milky Way galaxy. Its radio signal was intense enough that if the human eye were sensitive to only radio waves, then the galactic center would be among the brightest sources in the sky.

With some cleverly designed electronics, you can transmit specially encoded radio waves that can then be transformed into sound. This ingenious apparatus has come to be known as a "radio." So by virtue of extending our sense of sight, we have also, in effect, managed to extend our sense of hearing. But any source of radio waves, or practically any source of energy at all, can be channeled to vibrate the cone of a speaker, although journalists occasionally misunderstand this simple fact. For example, when radio emission was discovered from Saturn, it was simple enough for astronomers to hook up a radio receiver that was equipped with a speaker. The radio-wave signal was then converted to audible sound waves whereupon one journalist reported that "sounds" were coming from Saturn and that life on Saturn was trying to tell us something.

With much more sensitive and sophisticated radio detectors than were available to Karl Jansky, we now explore not just the Milky Way but the entire universe. As a testament to our initial seeing-is-believing bias, early detections of radio sources in the universe were often considered untrustworthy until they were confirmed by observations with a conventional telescope. Fortunately, most classes of radio-emitting objects also emit some level of visible light, so blind faith was not always required. Eventually, radio-wave telescopes produced a rich parade of discoveries that includes the still-mysterious quasars (a loosely assembled acronym of "quasi-stellar radio source"), which are among the most distant objects in the known universe.

Gas-rich galaxies emit radio waves from the abundant hydrogen atoms that are present (over 90 percent of all atoms in the universe are hydrogen). With large arrays of electronically connected radio telescopes we can generate very high resolution images of a galaxy's gas content that reveal intricate features in the hydrogen gas such as twists, blobs, holes, and filaments. In many ways the task of mapping galaxies is no different from that facing the fifteenth- and sixteenth-century cartographers, whose renditions of continents—distorted though they were—represented a noble human attempt to describe worlds beyond one's physical reach.

IF THE HUMAN EYE were sensitive to microwaves, then this window of the spectrum would enable you to see the radar emitted by the radar gun from the highway patrol officer who hides in the bushes. And microwave-emitting telephone relay station towers would be ablaze with light. Note, however, that the inside of your microwave oven would look no different because the mesh embedded in the door reflects microwaves back into the cavity to prevent their escape. The vitreous humor of your peering eyeballs is thus protected from getting cooked along with your food.

Microwave telescopes were not actively used to study the universe until the late 1960s. They allow us to peer into cool, dense clouds of interstellar gas that ultimately collapse to form stars and planets. The heavy elements in these clouds readily assemble into complex molecules whose signature in the microwave part of the spectrum is unmistakable because of their match with identical molecules that exist on Earth.

Some cosmic molecules are familiar to the household:

NH_3 (ammonia)
H_2O (water)

While some are deadly:

CO (carbon monoxide)
HCN (hydrogen cyanide)

Some remind you of the hospital:

H_2CO (formaldehyde)
C_2H_5OH (ethyl alcohol)

And some don't remind you of anything:

N_2H+ (dinitrogen monohydride ion)
CHC_3CN (cyanodiacetylene)

Nearly 130 molecules are known, including glycine, which is an amino acid that is a building block for protein and thus for life as we know it.

Without a doubt, a microwave telescope made the most important single discovery in astrophysics. The leftover heat from the big bang origin of the universe has now cooled to a temperature of about three degrees on the absolute temperature scale. (As fully detailed later in this section, the absolute temperature scale quite reasonably sets the coldest possible temperature to zero degrees, so there are no negative temperatures. Absolute zero corresponds to about -460 degrees Fahrenheit, while 310 degrees absolute corresponds to room temperature.) In 1965, this big bang remnant was serendipitously measured in a Nobel Prize–winning observation conducted at Bell Telephone Laboratories by the physicists Arno Penzias and Robert Wilson. The remnant manifests itself as an omnipresent and omnidirectional ocean of light that is dominated by microwaves.

This discovery was, perhaps, serendipity at its finest. Penzias and Wilson humbly set out to find terrestrial sources that interfered with microwave communications, but what they found was compelling evidence for the big bang theory of the origin of the universe, which must be like fishing for a minnow and catching a blue whale.

MOVING FURTHER ALONG the electromagnetic spectrum we get to infrared light. Also invisible to humans, it is most familiar to fast-food fanatics whose French fries are kept warm with infrared lamps for hours before purchase. These lamps also emit visible light, but their active ingredient is an abundance of invisible infrared photons that the food readily absorbs. If the human retina were sensitive to infrared, then an ordinary household scene at night, with all lights out, would reveal all objects that sustain a temperature in excess of room temperature, such as the household iron (provided it was turned on), the metal that surrounds the pilot lights of a gas stove, the hot water pipes, and the exposed skin of any humans who stepped into the scene. Clearly this picture is not more enlightening than what you would see with visible light, but you could imagine one or two creative uses of such vision, such as looking at your home in the winter to spot where heat leaks from the windowpanes or roof.

As a child, I knew that at night, with the lights out, infrared vision would discover monsters hiding in the bedroom closet only if they were warm-blooded. But everybody knows that your average bedroom monster is reptilian and cold-blooded. Infrared vision would thus miss a bedroom monster completely because it would simply blend in with the walls and the door.

In the universe, the infrared window is most useful as a probe of dense clouds that contain stellar nurseries. Newly formed stars are often enshrouded by leftover gas and dust. These clouds absorb most of the visible light from their embedded stars and reradiate it in the infrared, rendering our visible light window quite useless. While visible light gets heavily absorbed by interstellar dust clouds, infrared moves through with only minimal attenuation, which is especially valuable for studies in the plane of our own Milky Way galaxy because this is where the obscuration of visible light from the Milky Way's stars is at its greatest. Back home, infrared satellite photographs of Earth's surface reveal, among other things, the paths of warm oceanic currents such as the North Atlantic Drift current that swirls 'round the British Isles (which are farther north than the entire state of Maine) and keeps them from becoming a major ski resort.

The energy emitted by the Sun, whose surface temperature is about 6,000 degrees absolute, includes plenty of infrared, but peaks in the visible part of the spectrum, as does the sensitivity of the human retina, which, if you have never thought about it, is why our sight is so useful in the daytime. If this spectrum match were not so, then we could rightly complain that some of our retinal sensitivity was wasted. We don't normally think of visible light as penetrating, but light passes mostly unhindered through glass and air. Ultraviolet, however, is summarily absorbed by ordinary glass, so glass windows would not be much different from brick windows if our eyes were sensitive to only ultraviolet.

Stars that are over three or four times hotter than the Sun are prodigious producers of ultraviolet light. Fortunately, these stars are also bright in the visible part of the spectrum so discovering them has not depended on access to ultraviolet telescopes. The ozone layer in our atmosphere absorbs most of the ultraviolet, x-rays, and gamma rays that impinge upon it, so a detailed analysis of these hottest stars can best be obtained from Earth orbit or beyond. These high-energy windows in the spectrum thus represent relatively young subdisciplines of astrophysics.

AS IF TO herald a new century of extended vision, the first Nobel Prize ever awarded in physics went to the German physicist Wilhelm C. Röntgen in 1901 for his discovery of x-rays. Both ultraviolet and x-rays in the universe can reveal the presence of one of the most exotic objects in the universe: black holes. Black holes emit no light—their gravity is too strong for even light to escape—so their existence must be inferred from the energy emitted by matter that might spiral onto its surface from a companion star. The scene resembles greatly what water looks like as it spirals down a toilet bowl. With temperatures over twenty times that of the Sun's surface, ultraviolet and x-rays are the predominant form of energy released by material just before it descends into the black hole.

The act of discovery does not require that you understand either in advance, or after the fact, what you have discovered. This happened with the microwave background radiation and it is happening now with gamma ray bursts. As we will see in Section 6, the gamma-ray window has revealed mysterious bursts of high-energy gamma rays that are scattered across the sky. Their discovery was made possible through the use of space-borne gamma-ray telescopes, yet their origin and cause remain unknown.

If we broaden the concept of vision to include the detection of subatomic particles then we get to use neutrinos. As we saw in Section 2, the elusive neutrino is a subatomic particle that forms every time a proton transforms into an ordinary neutron and positron, which is the antimatter partner to an electron. As obscure as the process sounds, it happens in the Sun's core about a hundred billion billion billion billion (10^{38}) times each second. Neutrinos then pass directly out of the Sun as if it weren't there at all. A neutrino "telescope" would allow a direct view of the Sun's core and its ongoing thermonuclear fusion, which no band from the electromagnetic spectrum can reveal. But neutrinos are extraordinarily difficult to capture because they hardly ever interact with matter, so an efficient and effective neutrino telescope is a distant dream, if not an impossibility.

The detection of gravity waves, another elusive window on the universe, would reveal catastrophic cosmic events. But as of this writing, gravity waves, predicted in Einstein's general theory of relativity of 1916 as ripples in the fabric of space and time, have never been detected from any source. Physicists at the California Institute of Technology are developing a specialized gravity-wave detector that consists of an L-shaped evacuated pipe with 2.5-mile-long arms housing laser beams. If a gravitational wave passes by, the light path in one arm will temporarily differ in length from that of the other arm by a tiny amount. The experiment is known as LIGO, the Laser Interferometer Gravitational-wave Observatory, and it will be sensitive enough to detect gravitational waves from colliding stars over 100 million light-years away. One can imagine a time in the future where gravitational events in the universe—collisions, explosions, and collapsed stars—are observed routinely this way. Indeed, we may one day open this window wide enough to see beyond the opaque wall of microwave background radiation to the beginning of time itself.

SEVENTEEN

COLORS OF THE COSMOS

O nly a few objects in Earth's nighttime sky are bright enough to trigger our retina's color-sensitive cones. The red planet Mars can do it. As does the blue supergiant star Rigel (Orion's right kneecap) and the red supergiant Betelgeuse (Orion's left armpit). But aside from these standouts, the pickings are slim. To the unaided eye, space is a dark and colorless place.

Not until you aim large telescopes does the universe show its true colors. Glowing objects, like stars, come in three basic colors: red, white, and blue—a cosmic fact that would have pleased the founding fathers. Interstellar gas clouds can take on practically any color at all, depending on which chemical elements are present, and depending on how you photograph them, whereas a star's color follows directly from its surface temperature: Cool stars are red. Tepid stars are white. Hot stars are blue. Very hot stars are still blue. How about very, very hot places, like the 15-million-degree center of the Sun? Blue. To an astrophysicist, red-hot foods and red-hot lovers both leave room for improvement. It's just that simple.

Or is it?

A conspiracy of astrophysical law and human physiology bars the existence of green stars. How about yellow stars? Some astronomy textbooks, many science-fiction stories, and nearly every person on the street, comprise the Sun-Is-Yellow movement. Professional photographers, however, would swear the Sun is blue; "daylight" film is color-balanced on the expectation that the light source (presumably the Sun) is strong in the blue. The old blue-dot flash cubes were just one example of the attempt to simulate the Sun's blue light for indoor shots when using daylight film. Loft artists would argue, however, that the Sun is pure white, offering them the most accurate view of their selected paint pigments.

No doubt the Sun acquires a yellow-orange patina near the dusty horizon during sunrise and sunset. But at high noon, when atmospheric scattering is at a minimum, the color yellow does not spring to mind. Indeed, light sources that are truly yellow make white things look yellow. So if the Sun were pure yellow, then snow would look yellow—whether or not it fell near fire hydrants.

TO AN ASTROPHYSICIST, "cool" objects have surface temperatures between 1,000 and 4,000 degrees Kelvin and are generally described as red. Yet the filament of a high-wattage incandescent lightbulb rarely exceeds 3,000 degrees Kelvin (tungsten melts at 3,680 degrees) and looks very white. Below about 1,000 degrees, objects become dramatically less luminous in the visible part of the spectrum. Cosmic orbs with these temperatures are failed stars. We call them brown dwarfs even though they are not brown and emit hardly any visible light at all.

While we are on the subject, black holes aren't really black. They actually evaporate, very slowly, by emitting small quantities of light from the edge of their event horizon in a process first described by the physicist Stephen Hawking. Depending on a black hole's mass, it can emit any form of light. The smaller black holes are, the faster they evaporate, ending their lives in a runaway flash of energy rich in gamma rays, as well as visible light.

MODERN SCIENTIFIC IMAGES shown on television, in magazines, and in books often use a false color palette. TV weather forecasters have gone all the way, denoting things like heavy rainfall with one color and lighter rainfall with another. When astrophysicists create images of cosmic objects, they typically assign an arbitrary sequence of colors to an image's range of brightness. The brightest part might be red and the dimmest parts blue. So the colors you see bear no relation at all to the actual colors of the object. As in meteorology, some of these images have color sequences that relate to other attributes, such as the object's chemical composition or temperature. And it's not uncommon to see an image of a spiral galaxy that has been color-coded for its rotation: the parts coming toward you are shades of blue while the parts moving away are shades of red. In this case, the assigned colors evoke the widely recognized blue and red Doppler shifts that reveal an object's motion.

For the map of the famous cosmic microwave background, some areas are hotter than average. And, as must be the case, some areas are cooler than average. The range spans about one one-hundred-thousandth of a degree. How do you display this fact? Make the hot spots blue, and the cold spots red, or vice versa. In either case, a very small fluctuation in temperature shows up as an obvious difference on the picture.

Sometimes the public sees a full-color image of a cosmic object that was photographed using invisible light such as infrared, or radio waves. In most of these cases, we have assigned three colors, usually red, green, and blue (or "RGB" for short) to three different regions within the band. From this exercise, a full-color image can be constructed as though we were born with the capacity to see colors in these otherwise invisible parts of the spectrum.

The lesson is that common colors in common parlance can mean very different things to scientists than they do to everybody else. For the occasions when astrophysicists choose to speak unambiguously, we do have tools and methods that quantify the exact color emitted or reflected by an object, avoiding the tastes of the image maker or the messy business of human color perception. But these methods are not public-friendly. They involve the logarithmic ratio of the flux emitted by an object as measured through multiple filters in a well-defined system corrected for the detector's sensitivity profile. (See, I told you it wasn't public-friendly.) When that ratio decreases, for example, the object is technically turning blue no matter what color it appears to be.

THE VAGARIES OF human color perception took their toll on the wealthy American astronomer and Mars fanatic Percival Lowell. During the late 1800s and early 1900s, he made quite detailed drawings of the Martian surface. To make such observations, you need steady dry air, which reduces smearing of the planet's light en route to your eyeball. In the arid air of Arizona, atop Mars Hill, Lowell founded the Lowell Observatory in 1894. The iron-rich, rusty surface of Mars looks red at any magnification, but Lowell also recorded many patches of green at the intersections of what he described and illustrated as canals—artificial waterways, presumably made by real live Martians who were eager to distribute precious water from the polar icecaps to their cities, hamlets, and surrounding farmlands.

Let's not worry here about Lowell's alien voyeurism. Instead, let's just focus on his canals and green patches of vegetation. Percival was the unwitting victim of two well-known optical illusions. First, in almost all circumstances, the brain attempts to create visual order where there is no order at all. The constellations in the sky are prime examples—the result of imaginative, sleepy people asserting order on a random assortment of stars. Likewise, Lowell's brain interpreted uncorrelated surface and atmospheric features on Mars as large-scale patterns.

The second illusion is that gray, when viewed next to yellow-red, appears green-blue, an effect first pointed out by the French chemist M. E. Chevreul in 1839. Mars displays a dull red on its surface with regions of gray-brown. The green-blue arises from a physiological effect in which a color-neutral area surrounded by a yellow-orange appears bluish green to the eye.

In another peculiar but less embarrassing physiological effect, your brain tends to color balance the lighting environment in which you are immersed. Under the canopy of a rain forest, for example, where nearly all of the light that reaches the jungle floor has been filtered green (for having passed through leaves), a milk-white sheet of paper ought to look green. But it doesn't. Your brain makes it white in spite of the lighting conditions.

In a more common example, walk past a window at night while the people inside are watching television. If the TV is the only light in the room, the walls will glow a soft blue. But the brains of the people immersed in the light of the television actively color balance their walls and see no such discoloration around them. This bit of physiological compensation may prevent residents of our first Martian colony from taking notice of the prevailing red of their landscape. Indeed, the first images sent back to Earth in 1976 from the *Viking* lander, though pale, were purposefully color-tinted to a deep red so that they would fulfill the visual expectations of the press.

AT MID-TWENTIETH CENTURY, the night sky was systematically photographed from a location just outside San Diego, California. This seminal database, known as the Palomar Observatory Sky Survey, served as the foundation for targeted, follow-up observations of the cosmos for an entire generation. The cosmic surveyors photographed the sky twice, using identical exposures in two different kinds of black-and-white Kodak film—one ultrasensitive to blue light, the other ultrasensitive to red. (Indeed the Kodak corporation had an entire division whose job it was to serve the photographic frontier of astronomers, whose collective needs helped to push Kodak's R&D to its limits.) If a celestial object piqued your interest, you'd be sure to look at both the red- and blue-sensitive images as a first indication of the quality of light it emits. For example, extremely red objects are bright on the red image but barely visible on the blue. This kind of information informed subsequent observing programs for the targeted object.

Although modestly sized compared with the largest ground-based telescopes, the 94-inch *Hubble Space Telescope* has taken spectacular color images of the cosmos. The most memorable of these photographs are part of the Hubble Heritage series that will secure the telescope's legacy in the hearts and minds of the public. What astrophysicists do to make color images will surprise most people. First, we use the same digital CCD technology found in household camcorders, except that we used it a decade before you did and our detectors are much, much higher quality. Second, we filter the light in any one of several dozen ways before it hits the CCD. For an ordinary color photo, we obtain three successive images of the object, seen through broadband red, green, and blue filters. In spite of their names, taken together these filters span the entire visible spectrum. Next, we combine the three images in software the way the wetware of your brain combines the signals from the red-, green-, and blue-sensitive cones in your retina. This generates a color picture that greatly resembles what you would see if the iris in your eyeball were 94 inches in diameter.

Suppose, however, that the object were emitting light strongly at specific wavelengths due to the quantum properties of its atoms and molecules. If we know this in advance, and use filters tuned to these emissions, we can narrow our image sensitivity to just these wavelengths, instead of using broadband RGB. The result? Sharp features pop out of the picture, revealing structure and texture that would otherwise go unnoticed. A good example lives in our cosmic backyard. I confess to having never actually seen Jupiter's red spot through a telescope. While sometimes it's paler than at other times, the best way to see it is through a filter that isolates the red wavelengths of light coming from the molecules in the gas clouds.

In the galaxy, oxygen emits a pure green color when found near regions of star formation, amid the rarefied gas of the interstellar medium. (This was the mysterious element "nebulium" described earlier.) Filter for it and oxygen's signature arrives at the detector unpolluted by any ambient green light that may also occupy the scene. The vivid greens that jump out of many *Hubble* images come directly from oxygen's nighttime emissions. Filter for other atomic or molecular species and the color images become a chemical probe of the cosmos. The *Hubble* can do this so well that it's gallery of famous color images bears little resemblance to classical RGB images of the same objects taken by others who have tried to simulate the color response of the human eye.

The debate rages over whether or not these *Hubble* images contain "true" colors. One thing is certain, they do not contain "false" colors. They are the actual colors emitted by actual astrophysical objects and phenomena. Purists insist that we are doing a disservice to the public by not showing cosmic colors as the human eye would perceive them. I maintain, however, that if your retina were tunable to narrow-band light, then you would see just what the *Hubble* sees. I further maintain that my "if" in the previous sentence is no more contrived than the "if" in "If your eyes were the size of large telescopes."

The question remains, if you added together the visible light of all light-emitting objects in the universe, what color would you get? In simpler phrasing, What color is the universe? Fortunately, some people with nothing better to do have actually calculated the answer to this question. After an erroneous report that the universe is a cross between medium aquamarine and pale turquoise, Karl Glazebrook and Ivan Baldry of Johns Hopkins University corrected their calculations and determined that the universe is really a light shade of beige, or perhaps, cosmic latte. Glazebrook and Baldry's chromatic revelations came from a survey of the visible light from more than 200,000 galaxies, occupying a large and representative volume of the universe.

The nineteenth-century English astronomer Sir John Herschel invented color photography. To the frequent confusion but occasional delight of the public, astrophysicists have been messing with the process ever since—and will continue forever to do so.

EIGHTEEN

COSMIC PLASMA

Only in a few cases does a medical doctor's vocabulary overlap with that of the astrophysicist. The human skull has two "orbits" that shape the round cavities where our two eyeballs go; your "solar" plexus sits in the middle of your chest; and our eyes, of course, each have "lenses"; but our body has no quasars and no galaxies in it. For orbits and lenses, the medical and astrophysical usage resemble each other greatly. The term "plasma," however, is common to both disciplines, yet the two meanings have nothing whatever to do with each other. A transfusion of blood plasma can save your life, but a brief encounter with a glowing blob of million-degree astrophysical plasma would leave a puff of smoke where you had just been standing.

Astrophysical plasmas are remarkable for their ubiquity, yet they're hardly ever discussed in introductory textbooks or the popular press. In popular writings, plasmas are often called the fourth state of matter because of a panoply of properties that set them apart from familiar solids, liquids, and gases. A plasma has freely moving atoms and molecules, just like a gas, but a plasma can conduct electricity as well as lock onto magnetic fields that pass through it. Most atoms within a plasma have had electrons stripped from them by one mechanism or another. And the combination of high temperature and low density is such that the electrons only occasionally recombine with their host atoms. Taken as a whole, the plasma remains electrically neutral because the total number of (negatively charged) electrons equals the total number of (positively charged) protons. But inside, plasma seethes with electrical currents and magnetic fields and so, in many ways, behaves nothing like the ideal gas we all learned about in high-school chemistry class.

THE EFFECTS OF electric and magnetic fields on matter almost always dwarf the effects of gravity. The electrical force of attraction between a proton and an electron is forty powers of 10 stronger than their gravitational attraction. So strong are electromagnetic forces that a child's magnet easily lifts a paper clip off a tabletop in spite of Earth's formidable gravitational tug. Want a more interesting example? If you managed to extricate all the electrons from a cubic millimeter of atoms in the nose of the space shuttle, and if you affixed them all to the base of the launchpad, then the attractive force would inhibit the launch. All engines would fire and the shuttle wouldn't budge. And if the Apollo astronauts had brought back to Earth all electrons from a thimbleful of lunar dust (while leaving behind on the Moon the atoms from which they came), then their force of attraction would exceed the gravitational attraction between Earth and the Moon in its orbit.

The most conspicuous plasmas on Earth are fire, lightning, the trail of a shooting star, and of course, the electric shock you get after you shuffle around on your living room carpet in your wool socks and then touch a doorknob. Electrical discharges are jagged columns of electrons that abruptly move through the air when too many of them collect in one place. Across all the world's thunderstorms, Earth gets struck by lightning thousands of times per hour. The centimeter-wide air column through which a bolt of lightning travels becomes plasma in a fraction of a second as it is rendered aglow, having been raised to millions of degrees by these flowing electrons.

Every shooting star is a tiny particle of interplanetary debris moving so fast that it burns up in the air, harmlessly descending to Earth as cosmic dust. Almost the same thing happens to spacecraft that reenter the atmosphere. Since their occupants don't want to land at their orbital speed of 18,000 miles per hour (about five miles per second), the kinetic energy must go somewhere. It turns into heat on the leading edge of the craft during reentry and is rapidly whisked away by the heat shields. In this way, unlike shooting stars, the astronauts do not descend to Earth as dust. For several minutes during the descent, the heat is so intense that every molecule surrounding the space capsule becomes ionized, cloaking the astronauts in a temporary plasma barrier, through which none of our communication signals can penetrate. This is the famous blackout period when the craft is aglow and Mission Control knows nothing of the astronauts' well-being. As the craft continues to slow down through the atmosphere, the temperature cools, the air gets denser, and the plasma state can no longer be sustained. The electrons go back home to their atoms and communications are quickly restored.

WHILE RELATIVELY RARE on Earth, plasmas comprise more than 99.99 percent of all the visible matter in the cosmos. This tally includes all stars and gas clouds that are aglow. Nearly all of the beautiful photographs taken by the *Hubble Space Telescope* of nebulae in our galaxy depict colorful gas clouds in the form of plasma. For some, their shape and density are strongly influenced by the presence of magnetic fields from nearby sources. The plasma can lock a magnetic field into place and torque or otherwise shape the field to its whims. This marriage of plasma and magnetic field is a major feature of the Sun's 11-year cycle of activity. The gas near the Sun's equator rotates slightly faster than the gas near its poles. This differential is bad news for the Sun's complexion. With the Sun's magnetic field locked into its plasma, the field gets stretched and twisted. Sunspots, flares, prominences, and other solar blemishes come and go as the gnarly magnetic field punches through the Sun's surface, carrying solar plasma along with it.

Because of all the entanglement, the Sun flings up to a million tons per second of charged particles into space, including electrons, protons, and bare helium nuclei. This particle stream—sometimes a gale and sometimes a zephyr—is more commonly known as the solar wind. This most famous of plasmas is responsible for ensuring that comet tails point away from the Sun, no matter if the comet is coming or going. By colliding with molecules in Earth's atmosphere near our magnetic poles, the solar wind is also the direct cause of auroras (the northern and southern lights), not only on Earth but on all planets with atmospheres and strong magnetic fields. Depending on a plasma's temperature and its mix of atomic or molecular species, some free electrons will recombine with needy atoms and cascade down the myriad energy levels within. En route, the electrons emit light of prescribed wavelengths. Auroras owe their beautiful colors to these electron hijinks, as do neon tubes, fluorescent lights, as well as those glowing plasma spheres offered for sale next to the lava lamps in tacky gift shops.

These days, satellite observatories give us an unprecedented capacity to monitor the Sun and report on the solar wind as though it were part of the day's weather forecast. My first-ever televised interview for the evening news was triggered by the report of a plasma pie hurled by the Sun directly at Earth. Everybody (or at least the reporters) was scared that bad things would happen to civilization when it hit. I told the viewers not to worry—that we are protected by our magnetic field—and I invited them to use the occasion to go north and enjoy the aurora that the solar wind would cause.

THE SUN'S RAREFIED corona, visible during total solar eclipses as a glowing halo around the silhouetted near side of the Moon, is a 5-million-degree plasma that is the outermost part of the solar

atmosphere. With temperatures that high, the corona is the principal source of x-rays from the Sun, but is not otherwise visible to the
human eyes. Using visible light alone, the brightness of the Sun's
surface dwarfs that of the corona, easily getting lost in the glare.

There's an entire layer of Earth's atmosphere where electrons
have been kicked out of their host atoms by the solar wind, creating
a nearby blanket of plasma we call the ionosphere. This layer
reflects certain frequencies of radio waves, including those of the
AM dial on your radio. Because of this property of the ionosphere,
AM radio signals can reach hundreds of miles while "short wave"
radio can reach thousands of miles beyond the horizon. FM signals
and those of broadcast television, however, have much higher frequencies and pass right through, traveling out to space at the speed
of light. Any eavesdropping alien civilization will know all about
our TV programs (probably a bad thing), will hear all our FM
music (probably a good thing), and know nothing of the politics of
AM talk-show hosts (probably a safe thing).

Most plasmas are not friendly to organic matter. The person with
the most hazardous job on the *Star Trek* television series is the one
who must investigate the glowing blobs of plasma on the uncharted
planets they visit. (My memory tells me that this person always
wore a red shirt.) Every time this crew member meets a plasma
blob, he gets vaporized. Born of the twenty-fifth century, you'd
think these space-faring, star-trekking people would have long ago
learned to treat plasma with respect (or to not wear red). We in the
twenty-first century treat plasma with respect and we haven't been
anywhere.

IN THE CENTER of our thermonuclear fusion reactors, where plasmas are viewed from safe distances, we attempt to bring together
hydrogen nuclei at high speeds and turn them into heavier helium
nuclei. In so doing, we liberate energy that could supply society's
need for electricity. Problem is, we haven't yet succeeded in getting

more energy out than we put in. To achieve such high collision speeds, the blob of hydrogen atoms must be raised to tens of millions of degrees. No hope for attached electrons here. At these temperatures they've all been stripped from their hydrogen atoms and roam free. How might you hold a glowing blob of hydrogen plasma at millions of degrees? In what container would you place it? Even microwave-safe Tupperware will not do. What you need is a bottle that will not melt, vaporize, or decompose. As we saw briefly in Section 2, we can use the relationship between plasma and magnetic fields to our advantage and design a sort of "bottle" whose walls are intense magnetic fields that the plasma cannot cross. The economic return from a successful fusion reactor will rest in part on the design of this magnetic bottle and on our understanding of how the plasma interacts with it.

Among the most exotic forms of matter ever concocted is the newly isolated quark-gluon plasma, created by physicists at the Brookhaven National Laboratories, a particle accelerator facility on New York's Long Island. Rather than being filled with atoms stripped of their electrons, a quark-gluon plasma comprises a mixture of the most basic constituents of matter, the fractionally charged quarks and the gluons that normally hold them together to make protons and neutrons themselves. This unusual form of plasma resembles greatly the state of the entire universe, a fraction of a second after the big bang. This was about the time the observable universe could still fit within the 87-foot sphere of the Rose Center for Earth and Space. Indeed, in one form or another, every cubic inch of the universe was in a plasma state until nearly 400,000 years had elapsed.

Until then, the universe had cooled from trillions of degrees down to a few thousand. The whole time, all light was scattered left and right by the free electrons of our plasma-filled universe—a state that greatly resembles what happens to light as it passes through frosted glass or the Sun's interior. Light can travel through neither without scattering, rendering them both translucent instead of transparent. Below a few thousand degrees, the universe cooled enough for every electron in the cosmos to combine with one atomic nuclei, creating complete atoms of hydrogen and helium.

The pervasive plasma state no longer existed as soon as every electron found a home. And that's the way it would stay for hundreds of millions of years, at least until quasars were born, with their central black holes that dine on swirling gases. Just before the gas falls in, it releases ionizing ultraviolet light that travels across the universe, kicking electrons back out of their atoms with abandon. Until the quasars were born, the universe had enjoyed the only interval of time (before or since) where plasma was nowhere to be found. We call this era the Dark Ages and look upon it as a time when gravity was silently and invisibly assembling matter into plasma balls that became the first generation of stars.

NINETEEN

FIRE AND ICE

W hen Cole Porter composed "Too Darn Hot" for his 1948 Broadway musical *Kiss Me Kate*, the temperature he was bemoaning was surely no higher than the mid-nineties. No harm in taking Porter's lyrics as an authoritative source on the upper temperature limit for comfortable lovemaking. Combine that with what a cold shower does to most people's erotic urges, and you now have a pretty good estimate of how narrow the comfort zone is for the unclothed human body: a range of about 30 degrees Fahrenheit, with room temperature just about in the middle.

The universe is a whole other story. How does a temperature of 100,000,000,000,000,000,000,000,000,000,000 degrees grab you? That's a hundred thousand billion billion billion degrees. It also happens to be the temperature of the universe a teeny fraction of a second after the big bang—a time when all the energy and matter and space that would turn into planets, petunias, and particle physicists was an expanding fiery ball of quark-gluon plasma. Nothing you'd call a thing could exist until there was a multibillion-fold cooling of the cosmos.

As the laws of thermodynamics decree, within about one second after the big bang, the expanding fireball had cooled to 10 billion degrees and ballooned from something smaller than an atom to a cosmic colossus about a thousand times the size of our solar system. By the time three minutes had passed, the universe was a balmy billion degrees and was already hard at work making the simplest atomic nuclei. Expansion is the handmaiden to cooling, and the two have continued, unabated, ever since.

Today the average temperature of the universe is 2.73 degrees Kelvin. All the temperatures mentioned so far, aside from the ones that involve the human libido, are stated in degrees Kelvin. The Kelvin degree, known simply as the kelvin, was conceived to be the same temperature interval as the Celsius degree, but the Kelvin scale has no negative numbers. Zero is zero, period. In fact, to quash all doubts, zero on the Kelvin scale is dubbed absolute zero.

The Scottish engineer and physicist William Thomson, later and better known as Lord Kelvin, first articulated the idea of a coldest possible temperature in 1848. Laboratory experiments haven't gotten there yet. As a matter of principle, they never will, although they've come awfully close. The unarguably cold temperature of 0.0000000005 K (or 500 picokelvins, as metric mavens would say) was artfully achieved in 2003 in the lab of Wolfgang Ketterle, a physicist at MIT.

Outside the laboratory, cosmic phenomena span a staggering range of temperatures. Among the hottest places in the universe today is the core of a blue supergiant star during the hours of its collapse. Just before it explodes as a supernova, creating drastic neighborhood-warming effects, its temperature hits 100 billion K. Compare that with the Sun's core: a mere 15 million K.

Surfaces are much cooler. The skin of a blue supergiant checks in at about 25,000 K—hot enough, of course, to glow blue. Our Sun registers 6,000 K—hot enough to glow white, and hot enough to melt and then vaporize anything in the periodic table of elements. The surface of Venus is 740 K, hot enough to fry the electronics normally used to drive space probes.

Considerably further down the scale is the freezing point of water, 273.15 K, which looks downright warm compared with the 60 K surface of Neptune, nearly 3 billion miles from the Sun. Colder still is Triton, one of Neptune's moons. Its icy nitrogen surface sinks to 40 K, making it the coldest place in the solar system this side of Pluto.

Where do Earth-beings fit in? The average body temperature of humans (traditionally 98.6 degrees F) registers slightly above 310 on the Kelvin scale. Officially recorded surface temperatures on Earth range from a summer high of 331 K (136 F, at Al 'Aziziyah, Libya, in 1922) to a winter low of 184 K (−129 F, at Base Vostok, Antarctica, in 1983). But people can't survive unassisted at those extremes. We suffer hyperthermia in the Sahara if we don't have shelter from the heat, and hypothermia in the Arctic if we don't have boatloads of clothing and caravans of food. Meanwhile, Earth-dwelling extremophile microorganisms, both thermophilic (heat-loving) and psychrophilic (cold-loving), are variously adapted to temperatures that would fry us or freeze us. Viable yeast, wearing no clothes at all, has been discovered in 3-million-year-old Siberian permafrost. A species of bacterium locked in Alaskan permafrost for 32,000 years woke up and started swimming as soon as its medium melted. And at this very moment, assorted species of archaea and bacteria are living out their lives in boiling mud, bubbling hot springs, and undersea volcanoes.

Even complex organisms can survive in similarly astonishing circumstances. When provoked, the itsy-bitsy invertebrates known as tardigrades can suspend their metabolism. In that state, they can survive temperatures of 424 K (303 degrees F) for several minutes and 73 K (−328 degrees F) for days on end, making them hardy enough to endure being stranded on Neptune. So the next time you need space travelers with the "right stuff," you might want to choose yeast and tardigrades, and leave your astronauts, cosmonauts, and taikonauts* at home.

IT'S COMMON TO confuse temperature with heat. Heat is the total energy of all the motions of all the molecules in your substance of

*What the Chinese call their astronauts.

choice. It so happens that, within the mixture, the range of energies is large: some molecules move quickly, others move slowly. Temperature simply measures their average energy. For example, a cup of freshly brewed coffee may have a higher temperature than a heated swimming pool, but all the water in the pool holds vastly more heat than does the lone cup of coffee. If you rudely pour your 200-degree coffee into the 100-degree pool, the pool won't suddenly become 150 degrees. And whereas two people in a bed are a source of twice as much heat as one person in a bed, the average temperatures of their two bodies—98.6 and 98.6—do not normally add up to an undercover oven whose temperature is 197.2 degrees.

Scientists in the seventeenth and eighteenth centuries considered heat to be closely linked with combustion. And combustion, as they understood it, happened when phlogiston, a hypothetical earthlike substance characterized mainly by its combustibility, was removed from an object. Burn a log in the fireplace, air carries off the phlogiston, and the dephlogisticated log reveals itself as a pile of ashes.

By the late eighteenth century the French chemist Antoine-Laurent Lavoisier had replaced phlogiston theory with caloric theory. Lavoisier classified heat, which he called caloric, as one of the chemical elements, and contended that it was an invisible, tasteless, odorless, weightless fluid that passed between objects through combustion or rubbing. The concept of heat was not fully understood until the nineteenth century, the peak of the industrial revolution, when the broader concept of energy took shape within the new branch of physics called thermodynamics.

ALTHOUGH HEAT as a scientific idea posed plenty of challenges to brilliant minds, both scientists and nonscientists have intuitively grasped the concept of temperature for millennia. Hot things have a high temperature. Cold things have a low temperature. Thermometers confirm the connection.

Although Galileo is often credited with the invention of the thermometer, the earliest such device may have been built by the first-century A.D. inventor Heron of Alexandria. Heron's book *Pneumatica* includes a description of a "thermoscope," a device that showed the change in the volume of a gas as it was heated or cooled. Like many other ancient texts, *Pneumatica* was translated into Latin during the Renaissance. Galileo read it in 1594 and, as he later did when he learned of the newly invented telescope, he immediately constructed a better thermoscope. Several of his contemporaries did the same.

For a thermometer, scale is crucial. There's a curious tradition, beginning early in the eighteenth century, of calibrating the temperature units in such a way that common phenomena get assigned fraction-friendly numbers with many divisors. Isaac Newton proposed a scale from zero (melting snow) to 12 (the human body); 12 is, of course, evenly divisible by 2, 3, 4, and 6. The Danish astronomer Ole Rømer offered a scale from zero to 60 (60 being divisible by 2, 3, 4, 5, 6, 10, 12, 15, 20, and 30). On Rømer's scale, zero was the lowest temperature he could achieve with a mixture of ice, salt, and water; 60 was the boiling point of water.

In 1724 a German instrument maker named Daniel Gabriel Fahrenheit (who developed the mercury thermometer in 1714) came up with a more precise scale, splitting each degree of Rømer's into four equal parts. On the new scale, water boiled at 240 degrees and froze at 30, and human body temperature was about 90. After further adjustments, the span from zero to body temperature became 96 degrees, another winner in the divisibility department (its divisors are 2, 3, 4, 6, 8, 12, 16, 24, 32, and 48). The freezing point of water became 32 degrees. Still further tuning and standardization saddled fans of the Fahrenheit scale with a body temperature that isn't a round number, and a boiling point of 212 degrees.

Following a different path, in 1742 the Swedish astronomer Anders Celsius proposed a decimal-friendly centigrade scale for temperature. He set the freezing point at 100 and the boiling point at zero. This was not the first or last time an astronomer labeled a scale backward. Somebody, quite possibly the chap who manufactured Celsius's thermometers, did the world a favor and reversed the numbering, giving us the now-familiar Celsius scale. The number zero seems to have a crippling effect on some people's comprehension. One night a couple of decades ago, while I was on winter break from graduate school and was staying at my parents' house north of New York City, I turned on the radio to listen to classical music. A frigid Canadian air mass was advancing on the Northeast, and the announcer, between movements of George Frideric Handel's Water Music, continually tracked the descending outdoor temperature: "Five degrees Fahrenheit." "Four degrees." "Three degrees." Finally, sounding distressed, he announced, "If this keeps up, pretty soon there'll be no temperature left!"

In part to avoid such embarrassing examples of innumeracy, the international community of scientists uses the Kelvin temperature scale, which puts zero in the right place: at the absolute bottom. Any other location for zero is arbitrary and does not lend itself to play-by-play arithmetic commentary.

Several of Kelvin's predecessors, by measuring the shrinking volume of a gas as it cooled, had established −273.15 degrees Celsius (−459.67 degrees F) as the temperature at which the molecules of any substance have the least possible energy. Other experiments showed that −273.15 C is the temperature at which a gas, when kept at constant pressure, would drop to zero volume. Since there is no such thing as a gas with zero volume, −273.15 C became the unattainable lower limit of the Kelvin scale. And what better term to use for it than "absolute zero"?

THE UNIVERSE AS a whole acts somewhat like a gas. If you force a gas to expand, it cools. Back when the universe was a mere half-million years old, the cosmic temperature was about 3,000 K. Today it is less than 3 K. Inexorably expanding toward thermal oblivion, the present-day universe is a thousand times larger, and a thousand times cooler, than the infant universe.

On Earth, you normally measure temperatures by cramming a thermometer into a creature's orifice or letting the thermometer touch an object in some other, less intrusive way. This form of direct contact enables the moving molecules within the thermometer to reach the same average energy as the molecules in the object. When a thermometer sits idle in the air instead of performing its labors inside a rib roast, it's the average speed of the colliding air molecules that tell the thermometer what temperature to register.

Speaking of air, at a given time and place on Earth the air temperature in full sunlight is basically the same as the air temperature under a nearby tree. What the shade does is shield you from the Sun's radiant energy, nearly all of which passes unabsorbed through the atmosphere and lands on your skin, making you feel hotter than the air would by itself. But in empty space, where there is no air, there are no moving molecules to trigger a thermometer reading. So the question "What is the temperature of space?" has no obvious meaning. With nothing touching it, the thermometer can only register the radiant energy from all the light, from all sources, that lands upon it.

On the daytime side of our airless Moon, a thermometer would register 400 K (260 degrees F). Move a few feet into the shadow of a boulder, or journey to the Moon's night side, and the thermometer would instantly drop to 40 K (–390 degrees F). To survive a lunar day without wearing a temperature-controlled space suit, you would have to do pirouettes, alternately baking and then cooling all sides of your body, just to maintain a comfortable temperature.

WHEN THE GOING gets really cold and you want to absorb maximum radiant energy, wear something dark rather than reflective. The same holds for a thermometer. Rather than debate how to dress it in space, assume the thermometer can be made perfectly absorbent. If you now place it in the middle of nowhere, such as halfway between the Milky Way and the Andromeda galaxy, far from all obvious sources of radiation, the thermometer will settle at 2.73 K, the current background temperature of the universe.

A recent consensus among cosmologists holds that the universe will expand forever and ever. By the time the cosmos doubles in size, its temperature will drop by half. By the time it doubles again, its temperature will halve once more. With the passage of trillions of years, all the remaining gas will have been used to make stars, and all the stars will have exhausted their thermonuclear fuels. Meanwhile, the temperature of the expanding universe will continue to descend, approaching ever closer to absolute zero.

THE
MEANING
OF
LIFE

THE CHALLENGES AND TRIUMPHS
OF KNOWING HOW WE GOT HERE

TWENTY

DUST TO DUST

A casual look at the Milky Way with the unaided eye reveals a cloudy band of light and dark splotches extending from horizon to horizon. With the help of simple binoculars or a backyard telescope, the dark and boring areas of the Milky Way resolve into, well, dark and boring areas—but the bright areas resolve into countless stars and nebulae.

In a small book entitled *Sidereus Nuncius* (The Starry Messenger), published in Venice in 1610, Galileo gives an account of the heavens as seen through a telescope, including the first-ever description of the Milky Way's patches of light. Referring to his yet-to-be-named instrument as a "spyglass," he is so excited he can barely contain himself:

> *The Milky Way itself, which, with the aid of the spyglass, may be observed so well that all the disputes that for so many generations have vexed philosophers are destroyed by visible certainty, and we are liberated from wordy arguments. For the Galaxy is nothing else than a congeries of innumerable stars distributed in clusters. To whatever region of it you direct your spyglass, an immense number of stars immediately offer themselves to view, of which very many appear rather large and very conspicuous but the multitude of small ones is truly unfathomable.* (Van Helden 1989, p. 62)

Surely "immense number of stars" is where the action is. Why would anybody be interested in the dark areas where stars are absent? They are probably cosmic holes to the infinite and empty beyond.

Three centuries would pass before anybody figured out that the dark patches are thick, dense clouds of gas and dust, which obscure the more distant star fields and hold stellar nurseries deep within. Following earlier suppositions of the American astronomer George Cary Comstock, who wondered why faraway stars were much dimmer than their distance alone would indicate, it was not until 1909 when the Dutch astronomer Jacobus Cornelius Kapteyn (1851–1922) would name the culprit. In two research papers, both titled "On the Absorption of Light in Space," Kapteyn presented evidence that clouds, his newfound "interstellar medium," not only scatter the overall light of stars but do so unevenly across the rainbow of colors in a star's spectrum, attenuating the blue light more severely than the red. This selective absorption makes the Milky Way's faraway stars look, on average, redder than the near ones.

Ordinary hydrogen and helium, the principal constituents of cosmic gas clouds, don't redden light. But larger molecules do—especially those that contain the elements carbon and silicon. And when the molecules get too big to be called molecules, we call them dust.

MOST PEOPLE ARE familiar with dust of the household variety, although few know that, in a closed home, it consists mostly of dead, sloughed-off human skin cells (plus pet dander, if you own a live-in mammal). Last I checked, cosmic dust in the interstellar medium contains nobody's epidermis. But it does have a remarkable ensemble of complex molecules that emit principally in the infrared and microwave parts of the spectrum. Microwave telescopes were not a major part of the astrophysicist's tool kit until the 1960s; infrared telescopes, not until the 1970s. And so the true chemical richness of the stuff between the stars was unknown until then. In the decades that followed, a fascinating, intricate picture of star birth emerged.

Not all gas clouds in the Milky Way can form stars at all times. More often than not, the cloud is confused about what to do next. Actually, astrophysicists are the confused ones here. We know the cloud wants to collapse under its own weight to make one or more stars. But rotation as well as turbulent motion within the cloud work against that fate. So, too, does the ordinary gas pressure you learned about in high-school chemistry class. Galactic magnetic fields also fight collapse: they penetrate the cloud and latch onto any free-roaming charged particles contained therein, restricting the ways in which the cloud will respond to its self-gravity. The scary part is that if none of us knew in advance that stars exist, frontline research would offer plenty of convincing reasons for why stars could never form.

Like the Milky Way's several hundred billion stars, gas clouds orbit the center of the galaxy. The stars are tiny specks (a few light-seconds across) in a vast ocean of permeable space, and they pass one another like ships in the night. Gas clouds, on the other hand, are huge. Typically spanning hundreds of light-years, they contain the mass equivalent of a million Suns. As these clouds lumber through the galaxy, they often collide with one other, entangling their innards. Sometimes, depending on their relative speeds and their angles of impact, the clouds stick together like hot marshmallows; at other times, adding injury to insult, they rip each other apart.

If a cloud cools to a low enough temperature (less than about 100 degrees above absolute zero), its constituent atoms will bump and stick rather than careen off one another, as they do at higher temperatures. This chemical transition has consequences for everybody. The growing particles—now containing tens of atoms—begin to bat visible light to and fro, strongly attenuating the light of stars behind it. By the time the particles become full-grown dust grains, they contain upwards of 10 billion atoms. At that size, they no longer scatter the visible light from the stars behind them: they absorb it, then reradiate the energy as infrared, which is a part of the spectrum that freely escapes the cloud. But the act of absorbing visible light creates a pressure that pushes the cloud opposite the direction of the light source. The cloud is now coupled to starlight.

The forces that make the cloud more and more dense may eventually lead to its gravitational collapse, and that in turn leads to star birth. Thus we face an odd situation: to create a star with a 10-million-degree core, hot enough to undergo thermonuclear fusion, we must first achieve the coldest possible conditions within a cloud.

At this time in the life of a cloud, astrophysicists can only gesticulate what happens next. Theorists and computer modelers face the many parameter problem of inputting all known laws of physics and chemistry into their supercomputers before they can even think about tracking the dynamic behavior of large, massive clouds under all external and internal influences. A further challenge is the humbling fact that the original cloud is billions of times wider and a hundred sextillion times less dense than the star we're trying to create—and what matters on one size scale is not necessarily the right thing to worry about on another.

NEVERTHELESS, ONE THING we can safely assert is that in the deepest, darkest, densest regions of an interstellar cloud, with temperatures down around 10 degrees above absolute zero, pockets of gas do collapse without resistance, converting their gravitational energy into heat. The temperature in each region—soon to become the core of a newborn star—rises rapidly, dismantling all the dust grains in the immediate vicinity. Eventually the collapsing gas reaches 10 million degrees. At this magic temperature, protons (which are just naked hydrogen atoms) move fast enough to overcome their repulsion, and they bond under the influence of a short-range, strong nuclear force whose technical term is "strong nuclear force." This thermonuclear fusion creates helium, whose mass is less than the sum of its parts. The lost mass has been converted into boatloads of energy, as described by Einstein's famous equation $E = mc^2$, where E is energy, m is mass, and c is the speed of light. As the heat moves outward, the gas becomes luminous, and the energy that had formerly been mass now makes its exit. And although the region of hot gas still sits womblike within the greater cloud, we may nonetheless announce to the Milky Way that a star is born.

We know that stars come in a wide range of masses: from a mere one-tenth to nearly a hundred times that of the Sun. For reasons not yet divined, our giant gas cloud contains a multitude of cold pockets, all of which form at about the same time and each of which gives birth to a star. For every high-mass star born, there are a thousand low-mass stars. But only about 1 percent of all the gas in the original cloud participates in star birth, and that presents a classic challenge: figuring out how and why the tail wags the dog.

THE MASS LIMIT on the low end is easy to determine. Below about one-tenth of the Sun's mass, the pocket of collapsing gas does not have enough gravitational energy to bring its core temperature up to the requisite 10 million degrees. A star is not born. Instead we get what is commonly called a brown dwarf. With no energy source of its own, it just gets dimmer and dimmer over time, living off the little heat it was able to generate from its original collapse. The outer gaseous layers of a brown dwarf are so cool that many of the large molecules normally destroyed in the atmospheres of hotter stars remain alive and well within it. With such a feeble luminosity, a brown dwarf is supremely difficult to detect, requiring methods similar to those used for the detection of planets. Indeed, only in recent years have enough brown dwarfs been discovered to classify them into more than one category. The mass limit at the high end is also easy to determine. Above about a hundred times that of the Sun's mass, the star is so luminous that any additional mass that may want to join the star gets pushed away by the intense pressure of the star's light on the dust grains within the cloud, which carries the gas cloud with it. Here the coupling of starlight with dust is irreversible. So potent are the effects of this radiation pressure that the luminosity of just a few high-mass stars can disperse nearly all the mass from the original dark, obscuring cloud, thereby laying bare dozens, if not hundreds, of brand-new stars—siblings, really—for the rest of the galaxy to see.

The Great Nebula in Orion—situated just below Orion's belt, midway down his sword—is a stellar nursery of just that sort. Within the nebula thousands of stars are being born in one giant cluster. Four of the several massive ones form the Orion Trapezium and are busy evacuating a giant hole in the middle of the cloud from which they formed. New stars are clearly visible in *Hubble* telescope images of the region, each infant swaddled in a nascent, protoplanetary disk made of dust and other molecules drawn from the original cloud. And within each disk a solar system is forming.

For a long while, newborn stars don't bother anybody. But eventually, from the prolonged, steady gravitational perturbations of enormous passing clouds, the cluster ultimately falls apart, its members scattering into the general pool of stars in the galaxy. The low-mass stars live practically forever, so efficient is their consumption of fuel. The intermediate-mass stars, such as our Sun, sooner or later turn into red giants, expanding a hundredfold in size as they march toward death. Their outermost gaseous layers become so tenuously connected to the star that they drift into space, exposing the spent nuclear fuels that powered their 10-billion-year lives. The gas that returns to space gets swept up by passing clouds, only to participate in later rounds of the formation of stars.

In spite of the rarity of the highest-mass stars, they hold nearly all the evolutionary cards. They boast the highest luminosity (a million times that of the Sun) and, as a consequence, the shortest lives (only a few million years). And as we will shortly see, high-mass stars manufacture dozens of heavy elements, one after the other, starting with hydrogen and proceeding to helium, carbon, nitrogen, oxygen, and so forth, all the way to iron in their cores. They die spectacular deaths in supernova explosions, making yet more elements in their fires and briefly outshining their entire home galaxy. The explosive energy spreads the freshly minted elements across the galaxy, blowing holes in its distribution of gas and enriching nearby clouds with the raw materials to make dust of their own. The supernova-blast waves move supersonically through the clouds, compressing the gas and dust, and possibly creating pockets of very high density necessary to form stars in the first place.

As we will see in the next chapter, the supernova's greatest gift to the cosmos is to seed clouds with the heavy elements that form planets and protists and people, so that once again, further endowed by the chemical enrichment from a previous generation of high-mass stars, another star is born.

TWENTY-ONE

FORGED IN THE STARS

Not all scientific discoveries are made by lone, antisocial researchers. Nor are all discoveries accompanied by media headlines and best-selling books. Some involve many people, span many decades, require complicated mathematics, and are not easily summarized by the press. Such discoveries pass almost unnoticed by the general public.

My vote for the most underappreciated discovery of the twentieth century is the realization that supernovas—the explosive death throes of high-mass stars—are the primary source for the origin and relative mix of heavy elements in the universe. This unheralded discovery took the form of an extensive research paper published in 1957 in the journal *Reviews of Modern Physics* titled "The Synthesis of the Elements in Stars," by E. Margaret Burbidge, Geoffrey R. Burbidge, William Fowler, and Fred Hoyle. In the paper they built a theoretical and computational framework that freshly interpreted 40 years of musings by others on such hot topics as the sources of stellar energy and the transmutation of elements.

Cosmic nuclear chemistry is a messy business. It was messy in 1957 and it is messy now. The relevant questions have always included: How do the various elements from the famed periodic table of elements behave when subjected to assorted temperatures and pressures? Do the elements fuse or do they split? How easily is this accomplished? Does the process liberate or absorb energy?

The periodic table is, of course, much more than just a mysterious chart of a hundred, or so, boxes with cryptic symbols in them. It is a sequence of every known element in the universe arranged by increasing number of protons in their nuclei. The two lightest are hydrogen, with one proton, and helium, with two protons. Under the right conditions of temperature, density, and pressure, you can use hydrogen and helium to synthesize every other element on the periodic table.

A perennial problem in nuclear chemistry involves calculating accurate collision cross-sections, which are simply measures of how close one particle must get to another particle before they interact significantly. Collision cross-sections are easy to calculate for things such as cement mixers or houses moving down the street on flatbed trucks, but it can be a challenge for elusive subatomic particles. A detailed understanding of collision cross-sections is what enables you to predict nuclear reaction rates and pathways. Often small uncertainties in tables of collision cross-sections can force you to draw wildly erroneous conclusions. The problem greatly resembles what would happen if you tried to navigate your way around one city's subway system while using another city's subway map as your guide.

Apart from this ignorance, scientists had suspected for some time that if an exotic nuclear process existed anywhere in the universe, then the centers of stars were as good a place as any to find it. In particular, the British theoretical astrophysicist Sir Arthur Eddington published a paper in 1920 titled "The Internal Constitution of the Stars" where he argued that the Cavendish Laboratory in England, the most famous atomic and nuclear physics research center of the day, could not be the only place in the universe that managed to change some elements onto others:

> *But is it possible to admit that such a transmutation is occurring? It is difficult to assert, but perhaps more difficult to deny, that this is going on . . . and what is possible in the Cavendish Laboratory may not be too difficult in the sun. I think that the suspicion has been generally entertained that the stars are the crucibles in which the*

lighter atoms which abound in the nebulæ are compounded into more complex elements. (p. 18)

Eddington's paper predates by several years the discovery of quantum mechanics, without which our knowledge of the physics of atoms and nuclei was feeble, at best. With remarkable prescience, Eddington began to formulate a scenario for star-generated energy via the thermonuclear fusion of hydrogen to helium and beyond:

> We need not bind ourselves to the formation of helium from hydrogen as the sole reaction which supplies the energy [to a star], although it would seem that the further stages in building up the elements involve much less liberation, and sometimes even absorption, of energy. The position may be summarised in these terms: the atoms of all elements are built of hydrogen atoms bound together, and presumably have at one time been formed from hydrogen; the interior of a star seems as likely a place as any for the evolution to have occurred. (p. 18)

The observed mix of elements on Earth and elsewhere in the universe was another desirable thing for a model of the transmutation of the elements to explain. But first a mechanism was required. By 1931, quantum physics was developed (although the neutron was not yet discovered) and the astrophysicist Robert d'Escourt Atkinson published an extensive paper that he summarizes in his abstract as a "synthesis theory of stellar energy and of the origin of the elements . . . in which the various chemical elements are built up step by step from the lighter ones in stellar interiors, by the successive incorporation of protons and electrons one at a time" (p. 250).

At about the same time, the nuclear chemist William D. Harkins published a paper noting that "elements of low atomic weight are more abundant than those of high atomic weight and that, on the average, the elements with even atomic numbers are about 10 times more abundant than those with odd atomic numbers of similar value" (Lang and Gingerich 1979, p. 374). Harkins surmised that the relative abundances of the elements depend on nuclear rather than on conventional chemical processes and that the heavy elements must have been synthesized from the light ones.

The detailed mechanism of nuclear fusion in stars could ultimately explain the cosmic presence of many elements, especially those that you get each time you add the two-proton helium nucleus to your previously forged element. These constitute the abundant elements with "even atomic numbers" that Harkins refers to. But the existence and relative mix of many other elements remained unexplained. Another means of element buildup must have been at work.

The neutron, discovered in 1932 by the British physicist James Chadwick while working at the Cavendish Laboratory, plays a significant role in nuclear fusion that Eddington could not have imagined. To assemble protons requires hard work because they naturally repel each other. They must be brought close enough together (often by way of high temperatures, pressures, and densities) for the short-range "strong" nuclear force to overcome their repulsion and bind them. The chargeless neutron, however, repels no other particle, so it can just march into somebody else's nucleus and join the other assembled particles. This step has not yet created another element; by adding a neutron we have simply made an "isotope" of the original. But for some elements, the freshly captured neutron is unstable and it spontaneously converts itself into a proton (which stays put in the nucleus) and an electron (which escapes immediately). Like the Greek soldiers who managed to breach the walls of Troy by hiding inside the Trojan Horse, protons can effectively sneak into a nucleus under the guise of a neutron.

If the ambient flow of neutrons is high, then an atom's nucleus can absorb many in a row before the first one decays. These rapidly absorbed neutrons help to create an ensemble of elements that are identified with the process and differ from the assortment of elements that result from neutrons that are captured slowly.

The entire process is known as neutron capture and is responsible for creating many elements that are not otherwise formed by traditional thermonuclear fusion. The remaining elements in nature can be made by a few other means, including slamming high-energy light (gamma rays) into the nuclei of heavy atoms, which then break apart into smaller ones.

AT THE RISK of oversimplifying the life cycle of a high-mass star, it is sufficient to recognize that a star is in the business of making and releasing energy, which helps to support the star against gravity. Without it, the big ball of gas would simply collapse under its own weight. A star's core, after having converted its hydrogen supply into helium, will next fuse helium into carbon, then carbon to oxygen, oxygen to neon, and so forth up to iron. To successively fuse this sequence of heavier and heavier elements requires higher and higher temperatures for the nuclei to overcome their natural repulsion. Fortunately this happens naturally because at the end of each intermediate stage, the star's energy source temporarily shuts off, the inner regions collapse, the temperature rises, and the next pathway of fusion kicks in. But there is just one problem. The fusion of iron absorbs energy rather than releases it. This is very bad for the star because it can now no longer support itself against gravity. The star immediately collapses without resistance, which forces the temperature to rise so rapidly that a titanic explosion ensues as the star blows its guts to smithereens. During the explosion, the star's luminosity can increase a billionfold. We call them supernovas, although I always felt that the term "super-duper novas" would be more appropriate.

Throughout the supernova explosion, the availability of neutrons, protons, and energy enable elements to be created in many different ways. By combining (1) the well-tested tenets of quantum mechanics, (2) the physics of explosions, (3) the latest collision cross-sections, (4) the varied processes by which elements can transmutate into one another, and (5) the basics of stellar evolutionary theory, Burbidge, Burbidge, Fowler, and Hoyle decisively implicated supernova explosions as the primary source of all elements heavier than hydrogen and helium in the universe.

With supernovas as the smoking gun, they got to solve one other problem for free: when you forge elements heavier than hydrogen and helium inside stars, it does the rest of the universe no good unless those elements are somehow cast forth to interstellar space and made available to form planets and people. Yes, we are stardust.

I do not mean to imply that all of our cosmic chemical questions are solved. A curious contemporary mystery involves the element technetium, which, in 1937, was the first element to be synthesized in the laboratory. (The name technetium, along with other words that use the root prefix "tech-," derives from the Greek word *technetos*, which translates to "artificial.") The element has yet to be discovered naturally on Earth, but it has been found in the atmosphere of a small fraction of red giant stars in our galaxy. This alone would not be cause for alarm were it not for the fact that technetium has a half-life of a mere 2 million years, which is much, much shorter than the age and life expectancy of the stars in which it is found. In other words, the star cannot have been born with the stuff, for if it were, there would be none left by now. There is also no known mechanism to create technetium in a star's core *and* have it dredge itself up to the surface where it is observed, which has led to exotic theories that have yet to achieve consensus in the astrophysics community.

Red giants with peculiar chemical properties are rare, but nonetheless common enough for there to be a cadre of astrophysicists (mostly spectroscopists) who specialize in the subject. In fact, my professional research interests sufficiently overlap the subject for me to be a regular recipient of the internationally distributed *Newsletter of Chemically Peculiar Red Giant Stars* (not available on the newsstand). It typically contains conference news and updates on research in progress. To the interested scientist, these ongoing chemical mysteries are no less seductive than questions related to black holes, quasars, and the early universe. But you will hardly ever read about them. Why? Because once again, the media has predetermined what is not worthy of coverage, even when the news item is something as uninteresting as the cosmic origin of every element in your body.

SEND IN THE CLOUDS

For nearly all of the first 400 millennia after the birth of the universe, space was a hot stew of fast-moving, naked atomic nuclei with no electrons to call their own. The simplest chemical reactions were still just a distant dream, and the earliest stirrings of life on Earth lay 10 billion years in the future.

Ninety percent of the nuclei brewed by the big bang were hydrogen, most of the rest were helium, and a trifling fraction were lithium: the makings of the simplest elements. Not until the ambient temperature in the expanding universe had cooled from trillions down to about 3,000 degrees Kelvin did the nuclei capture electrons. In so doing, they turned themselves into legal atoms and introduced the possibility of chemistry. As the universe continued to grow bigger and cooler, the atoms gathered into ever larger structures—gas clouds in which the earliest molecules, hydrogen (H_2) and lithium hydride (LiH), assembled themselves from the earliest ingredients available in the universe. Those gas clouds spawned the first stars, whose masses were each about a hundred times that of our Sun. And at the core of each star raged a thermonuclear furnace, hell-bent on making chemical elements far heavier than the first and simplest three.

When those titanic first stars exhausted their fuel supplies, they blew themselves to smithereens and scattered their elemental entrails across the cosmos. Powered by the energy of their own explosions, they made yet heavier elements. Atom-rich clouds of gas, capable of ambitious chemistry, now gathered in space.

Fast forward to galaxies, the principal organizers of visible matter in the universe—and within them, gas clouds pre-enriched by the flotsam of the earliest exploding stars. Soon those galaxies would host generation after generation of exploding stars, and generation after generation of chemical enrichment—the wellspring of those cryptic little boxes that make up the periodic table of elements.

Absent this epic drama, life on Earth—or anywhere else—would simply not exist. The chemistry of life, indeed the chemistry of anything at all, requires that elements make molecules. Problem is, molecules don't get made, and can't survive, in thermonuclear furnaces or stellar explosions. They need a cooler, calmer environment. So how in the world did the universe get to be the molecule-rich place we now inhabit?

RETURN, FOR A MOMENT, to the element factory deep within a first-generation high-mass star.

As we just saw, there in the core, at temperatures in excess of 10 million degrees, fast-moving hydrogen nuclei (single protons) randomly slam into one another. The event spawns a series of nuclear reactions that, at the end of the day, yield mostly helium and a lot of energy. So long as the star is "on," the energy released by its nuclear reactions generates enough outward pressure to keep the star's enormous mass from collapsing under its own weight. Eventually, though, the star simply runs out of hydrogen fuel. What remains is a ball of helium, which just sits there with nothing to do. Poor helium. It demands a tenfold increase in temperature before it will fuse into heavier elements.

Lacking an energy source, the core collapses and, in so doing, heats up. At about 100 million degrees, the particles speed up and the helium nuclei finally fuse, slamming together fast enough to combine into heavier elements. When they fuse, the reaction releases enough energy to halt further collapse—at least for a while. Fused helium nuclei spend a bit of time as intermediate products (beryllium, for instance), but eventually three helium nuclei end up becoming a single carbon nucleus. (Much later, when carbon becomes a complete atom with its complement of electrons in place, it reigns as the most chemically fruitful atom in the periodic table.)

Meanwhile, back inside the star, fusion proceeds apace. Eventually the hot zone runs out of helium, leaving behind a ball of carbon surrounded by a shell of helium that is itself surrounded by the rest of the star. Now the core collapses again. When its temperature rises to about 600 million degrees, the carbon, too, starts slamming into its neighbors—fusing into heavier elements via more and more complex nuclear pathways, all the while giving off enough energy to stave off further collapse. The factory is now in full swing, making nitrogen, oxygen, sodium, magnesium, silicon.

Down the periodic table we go, until iron. The buck stops at iron, the final element to be fused in the core of first-generation stars. If you fuse iron, or anything heavier, the reaction absorbs energy instead of emitting it. But stars are in the business of making energy, so it's a bad day for a star when it finds itself staring at a ball of iron in its core. Without a source of energy to balance the inexorable force of its own gravity, the star's core swiftly collapses. Within seconds, the collapse and the attendant rapid rise in temperature trigger a monstrous explosion: a supernova. Now there's plenty of energy to make elements heavier than iron. In the explosion's aftermath, a vast cloud of all the elements inherited and manufactured by the star scatters into the stellar neighborhood. And consider the cloud's top ingredients: atoms of hydrogen, helium, oxygen, carbon, and nitrogen. Sound familiar? Except for helium, which is chemically inert, those elements are the main ingredients of life as we know it. Given the stunning variety of molecules those atoms can form, both with themselves and with others, they are also likely to be the ingredients of life as we *don't* know it.

The universe is now ready, willing, and able to form the first molecules in space and construct the next generation of stars.

IF GAS CLOUDS are to make enduring molecules, they must hold more than the right ingredients. They must also be cool. In clouds hotter than a few thousand degrees, the particles move too quickly—and so the atomic collisions are too energetic—to stick together and sustain molecules. Even if a couple of atoms manage to come together and make a molecule, another atom will shortly slam into them with enough energy to break them apart. The high temperatures and high-speed impacts that worked so well for fusion now work against chemistry.

Gas clouds can live long, happy lives as long as the turbulent motions of their inner pockets of gas hold them up. Occasionally, though, regions of a cloud slow down enough—and cool down enough—for gravity to win, causing the cloud to collapse. Indeed, the very process that forms molecules also serves to cool the cloud: when two atoms collide and stick, some of the energy that drove them together is captured in their newly formed bonds or emitted as radiation.

Cooling has a remarkable effect on a cloud's composition. Atoms now collide as if they were slow boats, sticking together and building molecules rather than destroying them. Because carbon readily binds with itself, carbon-based molecules can get large and complex. Some become physically entangled, like the dust that collects into dust bunnies under your bed. When the ingredients favor it, the same thing can happen with silicon-based molecules. In either case, each grain of dust becomes a happening place, studded with hospitable crevices and valleys where atoms can meet at their leisure and build even more molecules. The lower the temperature, the bigger and more complex the molecules can become.

AMONG THE EARLIEST and most common compounds to form—once the temperature drops below a few thousand degrees—are several familiar diatomic (two-atom) and triatomic (three-atom) molecules. Carbon monoxide (CO), for instance, stabilizes long before the carbon condenses into dust, and molecular hydrogen (H_2) becomes the prime constituent of cooling gas clouds, now sensibly called molecular clouds. Among the triatomic molecules that form next are water (H_2O), carbon dioxide (CO_2), hydrogen cyanide (HCN), hydrogen sulfide (H_2S), and sulfur dioxide (SO_2). There's also the highly reactive triatomic molecule H_3+, which is eager to feed its third proton to hungry neighbors, instigating further chemical trysts.

As the cloud continues to cool, dropping below 100 degrees Kelvin or so, bigger molecules arise, some of which may be lying around in your garage or kitchen: acetylene (C_2H_2), ammonia (NH_3), formaldehyde (H_2CO), methane (CH_4). In still cooler clouds you can find the chief ingredients of other important concoctions: antifreeze (made from ethylene glycol), liquor (ethyl alcohol), perfume (benzene), and sugar (glycoaldehyde), as well as formic acid, whose structure is similar to that of amino acids, the building blocks of proteins.

The current inventory of molecules drifting between the stars is heading toward 130. The largest and most structurally intricate of them are anthracene ($C_{14}H_{10}$) and pyrene ($C_{16}H_{10}$), discovered in 2003 in the Red Rectangle Nebula, about 2,300 light-years from Earth, by Adolf N. Witt of the University of Toledo in Ohio and his colleagues. Formed of interconnected, stable rings of carbon, anthracene and pyrene belong to a family of molecules that syllable-loving chemists call polycyclic aromatic hydrocarbons, or PAHs. And just as the most complex molecules in space are based on carbon, so, of course, are we.

THE EXISTENCE OF MOLECULES in free space, something now taken for granted, was largely unknown to astrophysicists before 1963—

remarkably late, considering the state of other sciences. The DNA molecule had already been described. The atom bomb, the hydrogen bomb, and ballistic missiles had all been "perfected." The Apollo program to land men on the Moon was in progress. Eleven elements heavier than uranium had been created in the laboratory.

This astrophysical shortfall came about because an entire window of the electromagnetic spectrum—microwaves—hadn't yet been opened. Turns out, as we saw in Section 3, the light absorbed and emitted by molecules typically falls in the microwave part of the spectrum, and so not until microwave telescopes came online in the 1960s was the molecular complexity of the universe revealed in all its splendor. Soon the murky regions of the Milky Way were shown to be churning chemical factories. Hydroxyl (OH) was detected in 1963, ammonia in 1968, water in 1969, carbon monoxide in 1970, ethyl alcohol in 1975—all mixed together in a gaseous cocktail in interstellar space. By the mid-1970s, the microwave signatures of nearly forty molecules had been found.

Molecules have a definite structure, but the electron bonds that hold the atoms together are not rigid: they jiggle and wiggle and twist and stretch. As it happens, microwaves have just the right range of energies to stimulate this activity. (That's why microwave ovens work: a bath of microwaves, at just the right energy, vibrates the water molecules in your food. Friction among those dancing particles generates heat, cooking the food rapidly from within.)

Just as with atoms, every species of molecule in space identifies itself by the unique pattern of features in its spectrum. That pattern can readily be compared with patterns catalogued in laboratories here on Earth; without the lab data, often supplemented by theoretical calculations, we wouldn't know what we were looking at. The bigger the molecule, the more bonds have been deputized to keep it together, and the more ways its bonds can jiggle and wiggle. Each kind of jiggling and wiggling has a characteristic spectral wavelength, or "color"; some molecules usurp hundreds or even thousands of "colors" across the microwave spectrum, wavelengths at which they either absorb or emit light when their electrons take a stretch. And extracting one molecule's signature from the rest of the signatures is hard work, sort of like picking out the sound of your toddler's voice in a roomful of screaming children during playtime. It's hard, but you can do it. All you need is an acute awareness of the kinds of sounds your kid makes. Therein is your laboratory template.

ONCE FORMED, a molecule does not necessarily lead a stable life. In regions where ferociously hot stars are born, the starlight includes copious amounts of UV, ultraviolet light. UV is bad for molecules because its high energy breaks the bonds between a molecule's constituent atoms. That's why UV is bad for you, too: it's always best to avoid things that decompose the molecules of your flesh. So forget that a gigantic gas cloud may be cool enough for molecules to form within it; if the neighborhood is bathed in UV, the molecules in the cloud are toast. And the bigger the molecule, the less it can withstand such an assault.

Some interstellar clouds are so big and dense, though, that their outer layers can shield their inner layers. UV gets stopped at the edge of town by molecules that give their lives to protect their brethren deep within, thereby retaining the complex chemistry that cold clouds enjoy.

But eventually the molecular Mardi Gras comes to an end. As soon as the center of the gas cloud—or any other pocket of gas—gets dense enough and cool enough, the average energy of the moving gas particles gets too weak to keep the structure from collapsing under its own weight. That spontaneous gravitational shrinkage pumps the temperature back up, turning the erstwhile gas cloud into a locus of blazing heat as thermonuclear fusion gets underway.

Yet another star is born.

INEVITABLY, INESCAPABLY, one might even say tragically, the chemical bonds—including all the organic molecules the cloud so diligently made en route to stardom—now break apart in the searing heat. The more diffuse regions of the gas cloud, however, escape this fate. Then there's the gas close enough to the star to be affected by its growing force of gravity, but not so close as to be pulled into the star itself. Within that cocoon of dusty gas, thick disks of condensing material enter a safe orbit around the star. And within those disks, old molecules can survive and new ones can form with abandon.

What we have now is a solar system in the making, soon to comprise molecule-rich planets and molecule-rich comets. Once there's some solid material, the sky's the limit. Molecules can get as fat as they like. Set carbon loose under those conditions, and you might even get the most complex chemistry we know. How complex? It goes by another name: biology.

TWENTY-THREE

GOLDILOCKS AND
THE THREE PLANETS

Once upon a time, some four billion years ago, the formation of the solar system was nearly complete. Venus had formed close enough to the Sun for the intense solar energy to vaporize what might have been its water supply. Mars formed far enough away for its water supply to be forever frozen. And there was only one planet, Earth, whose distance was "just right" for water to remain a liquid and whose surface would become a haven for life. This region around the Sun came to be known as the habitable zone.

Goldilocks (of fairy-tale fame) liked things "just right," too. One of the bowls of porridge in the Three Bears' cottage was too hot. Another was too cold. The third was just right, so she ate it. Also in the Three Bears's cottage, one bed was too hard. Another was too soft. The third was just right, so Goldilocks slept in it. When the Three Bears came home, they discovered not only missing porridge but also Goldilocks fast asleep in a bed. (I forget how the story ends, but if I were the Three Bears—omnivorous and at the top of the food chain—I would have eaten Goldilocks.)

The relative habitability of Venus, Earth, and Mars would intrigue Goldilocks, but the actual story of these planets is somewhat more complicated than three bowls of porridge. Four billion years ago leftover water-rich comets and mineral-rich asteroids were still pelting the planetary surfaces, although at a much slower rate than before. During this game of cosmic billiards, some planets had migrated inward from where they had formed while others were kicked up to larger orbits. And among the dozens of planets that had formed, some were on unstable orbits and crashed into the Sun or Jupiter. Others were ejected from the solar system altogether. In the end, the few that remained had orbits that were "just right" to survive billions of years.

Earth settled into an orbit with an average distance of 93 million miles from the Sun. At this distance, Earth intersects a measly one two-billionth of the total energy radiated by the Sun. If you assume that Earth absorbs all incident energy from the Sun, then our home planet's average is about 280 degrees Kelvin (50 degrees F), which falls midway between winter and summer temperatures. At normal atmospheric pressures, water freezes at 273 degrees and boils at 373 degrees Kelvin, so we are well-positioned for nearly all of Earth's water to remain in a happy liquid state.

Not so fast. Sometimes in science you can get the right answer for the wrong reasons. Earth actually absorbs only two-thirds of the energy that reaches it from the Sun. The rest is reflected back into space by Earth's surface (especially the oceans) and by the clouds. If reflectivity is factored into the equations, then the average temperature for Earth drops to about 255 degrees Kelvin, which is well below the freezing point of water. Something must be operating in modern times to raise our average temperature back to something a little more comfortable.

But wait once more. All theories of stellar evolution tell us that 4 billion years ago, when life was forming out of Earth's proverbial primordial soup, the Sun was a third less luminous than it is today, which would have placed Earth's average temperature even further below freezing.

Perhaps Earth in the distant past was simply closer to the Sun. But after the early period of heavy bombardment, no known mechanisms could have shifted stable orbits back and forth within the solar system. Perhaps the greenhouse effect was stronger in the past. We don't know for sure. What we do know is that habitable zones, as originally conceived, have only peripheral relevance to whether there may be life on a planet within them.

The famous Drake equation, invoked in the search for extraterrestrial intelligence, provides a simple estimate for the number of civilizations one might expect to find in the Milky Way galaxy. When the equation was conceived in the 1960s by the American astronomer Frank Drake, the concept of a habitable zone did not extend beyond the idea that there would be some planets at the "just right" distance from their host stars. A version of the Drake equation reads: Start with the number of stars in the galaxy (hundreds of billions). Multiply this large number by the fraction of stars with planets. Multiply what remains by the fraction of planets in the habitable zone. Multiply what remains by the fraction of those planets that evolved life. Multiply what remains by the fraction that have evolved intelligent life. Multiply what remains by the fraction that might have developed a technology with which to communicate across interstellar space. Finally, when you introduce a star formation rate and the expected lifetime of a technologically viable civilization you get the number of advanced civilizations that are out there now, possibly waiting for our phone call.

Small, cool, low-luminosity stars live for hundreds of billions and even possibly trillions of years, which ought to allow plenty of time for the planets around them to evolve a life-form or two, but their habitable zones fall very close to the host star. A planet that forms there will swiftly become tidally locked and always show the same face toward the star (just as the Moon always shows the same face to Earth) creating an extreme imbalance in planetary heating—all water on the planet's "near" side would evaporate while all water on the planet's "far" side would freeze. If Goldilocks lived there, we would find her eating oatmeal while turning in circles (like a rotisserie chicken) right on the border between eternal sunlight and eternal darkness. Another problem with the habitable zones around these long-lived stars is that they are extremely narrow; a planet in a random orbit is unlikely to find itself at a distance that is "just right."

Conversely, large, hot, luminous stars have enormous habitable zones in which to find their planets. Unfortunately these stars are rare, and live for only a few million years before they violently explode, so their planets make poor candidates in the search for life as we know it—unless, of course, some rapid evolution occurred. But animals that can do advanced calculus were probably not the first things to slither out of the primordial slime.

We might think of the Drake equation as Goldilocks mathematics—a method for exploring the chances of getting things just right. But the Drake equation as originally conceived misses Mars, which lies well beyond the habitable zone of the Sun. Mars displays countless meandering dry riverbeds, deltas, and floodplains, which constitute in-your-face evidence for running water in the Martian past.

How about Venus, Earth's "sister" planet? It falls smack dab within the Sun's habitable zone. Covered completely by a thick canopy of clouds, the planet has the highest reflectivity of any planet in the solar system. There is no obvious reason why Venus could not have been a comfortable place. But it happens to suffer from a monstrous greenhouse effect. Venus's thick atmosphere of carbon dioxide traps nearly 100 percent of the small quantities of radiation that reach its surface. At 750 degrees Kelvin (900°F) Venus is the hottest planet in the solar system, yet it orbits at nearly twice Mercury's distance from the Sun.

If Earth has sustained the continuous evolution of life through billions of years of storm and drama, then perhaps life itself provides a feedback mechanism that maintains liquid water. This notion was advanced by the biologists James Lovelock and Lynn Margulis in the 1970s and is referred to as the Gaia hypothesis. This influential, yet controversial idea requires that the mixture of species on Earth at any moment acts as a collective organism that continuously (yet unwittingly) tunes Earth's atmospheric composition and climate to promote the presence of life—and by implication, the presence of liquid water. I am intrigued by the idea. It has even become the darling of the New Age movement. But I'd bet there are some dead Martians and Venusians who advanced the same theory about their own planets a billion years ago.

THE CONCEPT OF a habitable zone, when broadened, simply requires an energy source of any variety to liquefy water. One of Jupiter's moons, icy Europa, is heated by the tidal forces of Jupiter's gravitational field. Like a racquetball that heats up after the continuous stress of getting hit, Europa is heated from the varying stress induced by Jupiter pulling more strongly on one side of the moon compared with the other. The consequence? Current observational and theoretical evidence suggest that below the kilometer-thick surface ice there is an ocean of liquid water, possibly slush. Given the fecundity of life within Earth's oceans, Europa remains the most tantalizing place in the solar system for the possibility of life outside Earth.

Another recent breakthrough in our concept of a habitable zone are the newly classified extremophiles, which are life-forms that not only exist but thrive in climactic extremes of hot and cold. If there were biologists among the extremophiles, they would surely classify themselves as normal and any life that thrived in room temperature as an extremophile. Among the extremophiles are the heat-loving thermophiles, commonly found at the midocean ridges, where pressurized water, superheated to well beyond its normal boiling point, spews out from below Earth's crust into the cold ocean basin. The conditions are not unlike those within a household pressure cooker, where high pressures are supplied by a heavy-duty pot with a lockable lid and the water is heated beyond ordinary boiling temperatures, without actually coming to a boil.

On the cold ocean floor, dissolved minerals instantly precipitate out from the hot water vents and form giant porous chimneys up to a dozen stories tall that are hot in their cores and cooler on their edges, where they make direct contact with the ocean water. Across this temperature gradient live countless life-forms that have never seen the Sun and couldn't care less if it were there. These hardy bugs live on geothermal energy, which is a combination of the leftover heat from Earth's formation and heat continuously leaching into Earth's crust from the radioactive decay of naturally occurring yet unstable isotopes of familiar chemical elements such as Aluminum-26, which lasts millions of years, and Potassium-40, which lasts billions.

At the ocean floor we have what may be the most stable ecosystem on Earth. What if a jumbo asteroid slammed into Earth and rendered all surface life extinct? The oceanic thermophiles would surely continue undaunted in their happy ways. They might even evolve to repopulate Earth's surface after each extinction episode. And what if the Sun were mysteriously plucked from the center of the solar system and Earth spun out of orbit, adrift in space? This event would surely not merit attention in the thermophile press. But in 5 billion years, the Sun will become a red giant as it expands to fill the inner solar system. Meanwhile, Earth's oceans will boil away and Earth, itself, will vaporize. Now that would be news.

If thermophiles are ubiquitous on Earth, we are lead to a profound question: Could there be life deep within all those rogue planets that were ejected from the solar system during its formation? These "geo"thermal reservoirs can last billions of years. How about the countless planets that were forcibly ejected by every other solar system that ever formed? Could interstellar space be teeming with life formed and evolved deep within these homeless planets? Far from being a tidy region around a star, receiving just the right amount of sunlight, the habitable zone is indeed everywhere. So the Three Bears's cottage was, perhaps, not a special place among fairy tales. Anybody's residence, even that of the Three Little Pigs, might contain a sitting bowl of food at a temperature that is just right. We have learned that the corresponding fraction in the Drake equation, the one that accounts for the existence of a planet within a habitable zone, may be as large as 100 percent.

What a hopeful fairy tale this is. Life, far from being rare and precious, may be as common as planets themselves.

And the thermophilic bacteria lived happily ever after—about 5 billion years.

TWENTY-FOUR

WATER, WATER

From the looks of some dry and unfriendly looking places in our solar system, you might think that water, while plentiful on Earth, is a rare commodity elsewhere in the galaxy. But of all molecules with three atoms, water is by far the most abundant. And in a ranking of the cosmic abundance of elements, water's constituents of hydrogen and oxygen are one and three in the list. So rather than ask why some places have water, we may learn more by asking why all places don't.

Starting in the solar system, if you seek a waterless, airless place to visit then you needn't look farther than Earth's Moon. Water swiftly evaporates in the Moon's near-zero atmospheric pressure and its two-week-long, 200-degree Fahrenheit days. During the two-week night, the temperature can drop to 250 degrees below zero, a condition that would freeze practically anything.

The Apollo astronauts brought with them, to and from the Moon, all the air and water (and air-conditioning) they needed for their round-trip journey. But missions in the distant future may not need to bring water or assorted products derived from it. Evidence from the *Clementine* lunar orbiter strongly supports a long-held contention that there may be frozen lakes lurking at the bottom of deep craters near the Moon's north and south poles. Assuming the Moon suffers an average number of impacts per year from interplanetary flotsam, then the mixture of impactors should include sizable water-rich comets. How big? The solar system contains plenty of comets that, when melted, could make a puddle the size of lake Erie.

While one wouldn't expect a freshly laid lake to survive many sun-baked lunar days at 200 degrees, any comet that happened to crash and vaporize will cast some of its water molecules in the bottom of deep craters near the poles. These molecules will sink into the lunar soils where they will remain forever because such places are the only places on the Moon where the "Sun don't shine." (If you otherwise thought the Moon had a perpetual dark side then you have been badly misled by many sources, no doubt including Pink Floyd's 1973 best-selling rock album *Dark Side of the Moon*.)

As light-starved Arctic and Antarctic dwellers know, the Sun never gets very high in the sky at any time of day or year. Now imagine living in the bottom of a crater whose rim was higher than the highest level the Sun ever reached. In such a crater on the Moon, where there is no air to scatter sunlight into shadows, you would live in eternal darkness.

ALTHOUGH ICE IN the cold and dark of your freezer evaporates over time (just look at cubes in your freezer's ice tray after you've come back from a long vacation), the bottoms of these craters are so cold that evaporation has effectively stopped for all needs of this discussion. No doubt about it, if we were ever to establish an outpost on the Moon it would benefit greatly from being located near such craters. Apart from the obvious advantages of having ice to melt, filter, then drink, you can also break apart the water's hydrogen from its oxygen. Use the hydrogen and some of the oxygen as active ingredients in rocket fuel and keep the rest of the oxygen for breathing. And in your spare time between space missions, you can always go ice skating on the frozen lake created with the extracted water.

Knowing that the Moon has been hit by impactors, as its pristine record of craters tells us, then one might expect Earth to have been hit too. Given Earth's larger size and stronger gravity, one might even expect us to have been hit many more times. It has been—from birth all the way to present day. In the beginning, Earth didn't just hatch from an interstellar void as a preformed spherical blob. It grew from the condensing protosolar gas cloud from which the other planets and the Sun were formed. Earth continued to grow by accreting small solid particles and eventually through incessant impacts with mineral-rich asteroids and water-rich comets. How incessant? The early impact rate of comets is suspected of being high enough to have delivered Earth's entire oceanic supply of water. But uncertainties (and controversies) remain. When compared with the water in Earth's oceans, the water in comets observed today is anomalously high in deuterium, a form of hydrogen that packs one extra neutron in its nucleus. If the oceans were delivered by comets, then the comets available to hit Earth during the early solar system must have had a somewhat different chemical profile.

And just when you thought it was safe to go outside, a recent study on the water level in Earth's upper atmosphere suggests that Earth regularly gets slammed by house-sized chunks of ice. These interplanetary snowballs swiftly vaporize on impact with the air, but they too contribute to Earth's water budget. If the observed rate has been constant over the 4.6 billion-year history of Earth, then these snowballs may also account for the world's oceans. When added to the water vapor that we know is out-gassed from volcanic eruptions, we have no shortage of ways that Earth could have acquired its supply of surface water.

Our mighty oceans now comprise over two-thirds of Earth's surface area, but only about one five-thousandth of Earth's total mass. While a small fraction of the total, the oceans weigh in at a hefty 1.5 quintillion tons, 2 percent of which is frozen at any given time. If Earth ever suffers a runaway greenhouse effect (like what has happened on Venus), then our atmosphere would trap excess amounts of solar energy, the air temperature would rise, and the oceans would swiftly evaporate into the atmosphere as they sustained a rolling boil. This would be bad. Apart from the obvious ways that Earth's flora and fauna will die, an especially pressing cause of death would result from Earth's atmosphere becoming three hundred times more massive as it thickens with water vapor. We would all be crushed.

Many features distinguish Venus from the other planets in the solar system, including its thick, dense, heavy atmosphere of carbon dioxide that imparts one hundred times the pressure of Earth's atmosphere. We would all get crushed there too. But my vote for Venus's most peculiar feature is the presence of craters that are all relatively young and uniformly distributed over its surface. This innocuous-sounding feature implicates a single planetwide catastrophe that reset the cratering clock by wiping out all evidence of previous impacts. A major erosive weather phenomenon such as a planetwide flood could do it. But so could widespread geologic (Venusiologic?) activity, such as lava flows, turning Venus's entire surface into the American automotive dream—a totally paved planet. Whatever reset the clock, it must have ceased abruptly. But questions remain. If indeed there was a planetwide flood on Venus, where is all the water now? Did it sink below the surface? Did it evaporate into the atmosphere? Or was the flood composed of a common substance other than water?

OUR PLANETARY FASCINATION (and ignorance) is not limited to Venus. With meandering riverbeds, floodplains, river deltas, networks of tributaries, and river-eroded canyons, Mars was once a watering hole. The evidence is strong enough to declare that if anyplace in the solar system other than Earth ever boasted a flourishing water supply, it was Mars. For reasons unknown, Mars's surface is today bone dry. Whenever I look at both Venus and Mars, our sister and brother planets, I look at Earth anew and wonder how fragile our surface supply of liquid water just might be.

As we already know, imaginative observations of the planet by Percival Lowell led him to suppose that colonies of resourceful Martians had built an elaborate network of canals to redistribute

water from Mars's polar ice caps to the more populated middle latitudes. To explain what he thought he saw, Lowell imagined a dying
civilization that was somehow running out of water. In his thorough, yet curiously misguided treatise *Mars as the Abode of Life*,
published in 1909, Lowell laments the imminent end of the
Martian civilization he imagined he saw:

> *The drying up of the planet is certain to proceed until its surface can*
> *support no life at all. Slowly but surely time will snuff it out. When*
> *the last ember is thus extinguished, the planet will roll a dead world*
> *through space, its evolutionary career forever ended.* (p. 216)

Lowell happened to get one thing right. If there were ever a civilization (or any kind of life at all) that required water on the
Martian surface, then at some unknown time in Martian history,
and for some unknown reason, all the surface water *did* dry up,
leading to the exact fate for life that Lowell describes. Mars's missing water may be underground, trapped in the planet's permafrost.
The evidence? Large craters on the Martian surface are more
likely than small craters to exhibit dried mud-spills over their
rims. Assuming the permafrost to be quite deep, reaching it would
require a large collision. The deposit of energy from such an
impact would melt this subsurface ice on contact, enabling it to
splash upward. Craters with this signature are more common in
the cold, polar latitudes—just where one might expect the permafrost layer to be closer to the Martian surface. By some estimates, if all the water suspected of hiding in the Martian
permafrost and known to be locked in the polar ice caps were
melted and spread evenly over its surface, Mars would don a planetwide ocean tens of meters deep. A thorough search for contemporary (or fossil) life on Mars must include a plan to look many
places, especially below the Martian surface.

When thinking about where liquid water might be found (and by
association, life), astrophysicists were originally inclined to consider planets that orbited the right distance from their host star to

keep water in liquid form—not too close and not too far. This Goldilocks-inspired habitable zone, as it came to be known, was a good start. But it neglected the possibility of life in places where other sources of energy may be responsible for keeping water as a liquid when it might have otherwise turned to ice. A mild greenhouse effect would do it. So would an internal source of energy such as leftover heat from the formation of the planet or the radioactive decay of unstable heavy elements, each of which contributes to Earth's residual heat and consequent geologic activity.

Another source of energy are planetary tides, a more general concept than simply the dance between a moon and a sloshing ocean. As we have seen, Jupiter's moon Io gets continually stressed by changing tides as it ambles slightly closer and then slightly farther from Jupiter during its near-circular orbit. With a distance from the Sun that would otherwise guarantee a forever-frozen world, Io's stress level earns it the title of the most geologically active place in the entire solar system—complete with belching volcanoes, surface fissures, and plate tectonics. Some have analogized modern-day Io to the early Earth, when our planet was still piping hot from its episode of formation.

An equally intriguing moon of Jupiter is Europa, which also happens to be tidally heated. As had been suspected for some time, Europa was recently confirmed (from images taken by the *Galileo* planetary probe) to be a world covered with thick, migrating ice sheets, afloat on a subsurface ocean of slush or liquid water. An ocean of water! Imagine going ice fishing there. Indeed, engineers and scientists at the Jet Propulsion Laboratory are beginning to think about a mission where a space probe lands, finds (or cuts or melts) a hole in the ice, and extends a submersible camera to have a peek. Since oceans were the likely place of origin for life on Earth, the existence of life in Europa's oceans becomes a plausible fantasy.

In my opinion, the most remarkable feature of water is not the well-earned badge of "universal solvent" that we all learned in chemistry class; nor is it the unusually wide temperature range over which it remains liquid. As we have already seen, water's most remarkable feature is that, while most things—water included—shrink and become denser as they cool, water expands when it cools below 4 degrees Celsius, becoming less and less dense. When water freezes at zero degrees, it becomes even less dense than at any temperature when it was liquid, which is bad news for drainage pipes, but very good news for fish. In the winter, as the outside air drops below freezing, 4-degree water sinks to the bottom and stays there while a floating layer of ice builds extremely slowly on the surface, insulating the warmer water below.

Without this density inversion below 4 degrees, whenever the outside air temperature fell below freezing, the upper surface of a bed of water would cool and sink to the bottom as warmer water rose from below. This forced convection would rapidly drop the water's temperature to zero degrees as the surface begins to freeze. The denser, solid ice would sink to the bottom and force the entire bed of water to freeze solid from the bottom up. In such a world, there would be no ice fishing because all the fish would be dead—fresh frozen. And ice anglers would find themselves sitting on a layer of ice that either was submerged below all remaining liquid water or was atop a completely frozen body of water. No longer would you need icebreakers to traverse the frozen Arctic—either the entire Arctic ocean would be frozen solid or the frozen parts would all have sunk to the bottom and you could just sail your ship without incident. You could walk around, fearless of falling through. In this altered world, ice cubes and icebergs would sink, and in 1912, the *Titanic* would have steamed safely into its port of call in New York City.

The existence of water in the galaxy is not limited to planets and their moons. Water molecules, along with several other household chemicals such as ammonia and methane and ethyl alcohol, are found routinely in cool interstellar gas clouds. Under special conditions of low temperature and high density, an ensemble of water molecules can be induced to transform and funnel energy from a nearby star into an amplified, high-intensity beam of microwaves. The atomic physics of this phenomenon greatly resembles what goes on with visible light inside a laser. But in this case, the relevant acronym is M-A-S-E-R, for microwave amplification by the stimulated emission of radiation. Not only is water practically everywhere in the galaxy, it occasionally beams at you, too.

While we know water to be essential for life on Earth, we can only presume it to be a prerequisite for life elsewhere in the galaxy. Among the chemically illiterate, however, water is a deadly substance to be avoided. A now-famous science fair experiment that tested antitechnology sentiments and associated chemical-phobia was conducted in 1997 by Nathan Zohner, a 14-year-old student at Eagle Rock Junior High School in Idaho. He invited people to sign a petition that demanded either strict control of, or a total ban on, dihydrogen monoxide. He listed some of the odious properties of this colorless and odorless substance:

- It is a major component in acid rain
- It eventually dissolves almost anything it comes in contact with
- It can kill if accidentally inhaled
- It can cause severe burns in its gaseous state
- It has been found in tumors of terminal cancer patients

Forty-three out of 50 people approached by Zohner signed the petition, six were undecided, and one was a great supporter of dihydrogen monoxide and refused to sign. Yes, 86 percent of the passersby voted to ban water (H_2O) from the environment.

Maybe that's what really happened to all the water on Mars.

TWENTY-FIVE

LIVING SPACE

I f you ask people where they're from, they will typically say the name of the city where they were born, or perhaps the place on Earth's surface where they spent their formative years. Nothing wrong with that. But an astrochemically richer answer might be, "I hail from the explosive jetsam of a multitude of high-mass stars that died more than 5 billion years ago."

Outer space is the ultimate chemical factory. The big bang started it all, endowing the universe with hydrogen, helium, and a smattering of lithium: the three lightest elements. Stars forged all the rest of the ninety-two naturally occurring elements, including every bit of carbon, calcium, and phosphorus in every living thing on Earth, human or otherwise. How useless this rich assortment of raw materials would be had it stayed locked up in the stars. But when stars die, they return much of their mass to the cosmos, sprinkling nearby gas clouds with a portfolio of atoms that enrich the next generation of stars.

Under the right conditions of temperature and pressure, many of the atoms join to form simple molecules. Then, through routes both intricate and inventive, many molecules grow larger and more complex. Eventually, in what must surely be countless billions of places in the universe, complex molecules assemble themselves into some kind of life. In at least one cosmic corner, the molecules have become so complex that they have achieved consciousness and attained the ability to formulate and communicate the ideas conveyed by the marks on this page.

Yes, not only humans but also every other organism in the cosmos, as well as the planets or moons on which they thrive, would not exist but for the wreckage of spent stars. So you're made of detritus. Get over it. Or better yet, celebrate it. After all, what nobler thought can one cherish than that the universe lives within us all?

TO COOK UP some life, you don't need rare ingredients. Consider the top five constituents of the cosmos, in order of their abundance: hydrogen, helium, oxygen, carbon, and nitrogen. Take away chemically inert helium—which is not fond of making molecules with anybody—and you've got the top four constituents of life on Earth. Awaiting their cue within the massive clouds that lurk among a galaxy's stars, these elements begin making molecules as soon as the temperature drops below a couple thousand degrees Kelvin.

Molecules made of just two atoms form early: carbon monoxide and the hydrogen molecule (hydrogen atoms bound together in pairs). Drop the temperature some more, and you get stable three- or four-atom molecules such as water (H_2O), carbon dioxide (CO_2), and ammonia (NH_3)—simple but top-shelf ingredients in the kitchen of life. Drop the temperature even more, and hordes of five- and six-atom molecules form. And because carbon is both abundant and chemically enterprising, most of the molecules include it; indeed, three-quarters of all molecular "species" sighted in interstellar space have at least one carbon atom.

Sounds promising. But space can be a dangerous place for molecules. If the energy from stellar explosions doesn't destroy them, ultraviolet light from nearby ultraluminous stars will. The bigger the molecule, the less stable it is against assault. Molecules lucky enough to inhabit uneventful or shielded neighborhoods may endure long enough to be incorporated into grains of cosmic dust, and ultimately into asteroids, comets, planets, and people. Yet even if none of the original molecules survives the stellar violence, plenty of atoms and time remain available to make complex molecules, not only during the formation of a particular planet but also on and within the planet's nubile surface. Notables on the short list of complex molecules include adenine (one of the nucleotides, or "bases," that make up DNA), glycine (a protein precursor), and glycoaldehyde (a carbohydrate). Such ingredients, and others of their caliber, are essential for life as we know it and are decidedly not unique to Earth.

BUT ORGIES OF organic molecules are not life, just as flour, water, yeast, and salt are not bread. Although the leap from raw ingredients to living individual remains mysterious, several prerequisites are clear. The environment must encourage molecules to experiment with one another and must shelter them from excessive harm as they do so. Liquids offer a particularly attractive environment, because they enable both close contact and great mobility. The more chemical opportunities an environment affords, the more imaginative its resident experiments can be. Another essential factor, brought to you by the laws of physics, is a generous supply of energy to drive chemical reactions.

Given the wide range of temperatures, pressures, acidity, and radiation flux at which life thrives on Earth, and knowing that one microbe's cozy nook can be another's house of torture, scientists cannot at present stipulate additional requirements for life elsewhere. As a demonstration of the limits of this exercise, we find the charming little book *Cosmotheoros*, by the seventeenth-century Dutch astronomer Christiaan Huygens, wherein the author speculates that life-forms on other planets must grow hemp, for how else would they weave ropes to steer their ships and sail the open seas?

Three centuries later, we're content with just a pile of molecules. Shake 'em and bake 'em, and within a few hundred million years you might have thriving colonies of organisms.

LIFE ON EARTH is astonishingly fertile, that's for sure. But what about the rest of the universe? If somewhere there's another celestial body that bears any resemblance to our own planet, it may have run similar experiments with its similar chemical ingredients, and those experiments would have been choreographed by the physical laws that hold sway throughout the universe.

Consider carbon. Its capacity to bind in multiple ways, both to itself and to other elements, gives it a chemical exuberance unequalled in the periodic table. Carbon makes more kinds of molecules (how does 10 million grab you?) than all other elements combined. A common way for atoms to make molecules is to share one or more of their outermost electrons, creating a mutual grip analogous to the fist-shaped coupler between freight cars. Each carbon atom can bind with one, two, three, or four other atoms in this way, whereas a hydrogen atom binds with only one, oxygen with one or two, and nitrogen with three.

By binding to itself, carbon can generate myriad combinations of long-chain, highly branched, or closed-ring molecules. Such complex organic molecules are ripe for doing things that small molecules can only dream about. They can, for example, perform one kind of task at one end and another kind at the other; they can coil and curl and intertwine with other molecules, creating no end of features and properties. Perhaps the ultimate carbon-based molecule is DNA: a double-stranded chain that encodes the identity of all life as we know it.

What about water? When it comes to fostering life, water has the highly useful property of staying liquid across what most biologists regard as a fairly wide range of temperatures. Trouble is, most biologists look to Earth, where water stays liquid across 100 degrees of the Celsius scale. But on some parts of Mars, atmospheric pressure is so low that water is never liquid: a freshly poured cup of H_2O boils and freezes at the same time! Yet in spite of Mars's current sorry state, its atmosphere once supported liquid water in abundance. If ever the Red Planet harbored life on its surface, it would have been then.

Earth, of course, happens to have a goodly—and occasionally deadly—amount of water on its surface. Where did it come from? As we saw earlier, comets are a logical source: they're chock full of (frozen) water, the solar system holds countless billions of them, some are quite large, and they would regularly have been slamming into the early Earth back when the solar system was forming. Another source of water could have been volcanic outgassing, a frequent phenomenon on the young Earth. Volcanoes erupt not simply because magma is hot, but because hot, rising magma turns underground water to steam, which then expands explosively. The steam no longer fits in its subterranean chamber, and so the volcano blows its lid, bringing H_2O to Earth's surface from below. All things considered, then, the presence of water on our planet's surface is hardly surprising.

ALTHOUGH EARTH-LIFE takes multifarious forms, all of it shares common stretches of DNA. The biologist who has Earth-on-the-brain may revel in life's diversity, but the astrobiologist dreams of diversity on a grander scale: life based on alien DNA, or on something else entirely. Sadly, our planet is a singular biological sample. Nevertheless, the astrobiologist may glean insights about life-forms that dwell elsewhere in the cosmos by studying organisms that thrive in extreme environments here on Earth.

Once you look for them, you find these extremophiles practically everywhere: nuclear dump sites, acid-laden geysers, iron-saturated acidic rivers, chemical-belching vents on the ocean floor, submarine volcanoes, permafrost, slag heaps, commercial salt-evaporation ponds, and a host of other places you would not elect to spend your honeymoon but that may be more typical of the rest of the planets and moons out there. Biologists once presumed that life began in "some warm little pond," to quote Darwin (1959, p. 202); in recent years, though, the weight of evidence has tilted in favor of the view that extremophiles were the earliest earthly life-forms.

As we will see in the next section, for its first half-billion years, the inner solar system resembled a shooting gallery. Earth's surface was continually pulverized by crater-forming boulders large and small. Any attempt to jump-start life would have been swiftly aborted. By about 4 billion years ago, though, the impact rate slowed and Earth's surface temperature began to drop, permitting experiments in complex chemistry to survive and thrive. Older textbooks start their clocks at the birth of the solar system and typically declare that life on Earth needed 700 million or 800 million years to form. But that's not fair: the planet's chem-lab experiments couldn't even have begun until the aerial bombardment lightened up. Subtract 600 million years' worth of impacts right off the top, and you've got single-celled organisms emerging from the primordial ooze within a mere 200 million years. Even though scientists continue to be stumped about how life began, nature clearly had no trouble creating the stuff.

IN JUST A FEW dozen years, astrochemists have gone from knowing nothing of molecules in space to finding a plethora of them practically everywhere. Moreover, in the past decade astrophysicists have confirmed that planets orbit other stars and that every exosolar star system is laden with the same top four ingredients of life as our own cosmic home is. Although no one expects to find life on a star, even a thousand-degree "cool" one, Earth has plenty of life in places that register several hundred degrees. Taken together, these discoveries suggest it's reasonable to think of the universe as fundamentally familiar rather than as utterly alien.

But how familiar? Are all life-forms likely to be like Earth's—carbon-based and committed to water as their favorite fluid?

Take silicon, one of the top ten elements in the universe. In the periodic table, silicon sits directly below carbon, indicating that they have an identical configuration of electrons in their outer shells. Like carbon, silicon can bind with one, two, three, or four other atoms. Under the right conditions, it can also make long-chain molecules. Since silicon offers chemical opportunities similar to those of carbon, why couldn't life be based on silicon?

One problem with silicon—apart from its being a tenth as abundant as carbon—is the strong bonds it creates. When you link silicon and oxygen, for instance, you don't get the seeds of organic chemistry; you get rocks. On Earth, that's chemistry with a long shelf life. For chemistry that's friendly to organisms, you need bonds that are strong enough to survive mild assaults on the local environment but not so strong that they don't allow further experiments to take place.

And how important is liquid water? Is it the only medium suitable for chemistry experiments—the only medium that can shuttle nutrients from one part of an organism to another? Maybe life just needs a liquid. Ammonia is common. So is ethanol. Both are drawn from the most abundant ingredients in the universe. Ammonia mixed with water has a vastly lower freezing point (around −100 degrees Fahrenheit) than does water by itself (32 degrees), broadening the conditions under which you might find liquid-loving life. Or here's another possibility: on a world that lacks an internal heat source, orbits far from its host star, and is altogether bone-cold, normally gaseous methane might become the liquid of choice.

IN 2005, the European Space Agency's *Huygens* probe (named after you-know-who) landed on Saturn's largest moon, Titan, which hosts lots of organic chemistry and supports an atmosphere ten times thicker than Earth's. Setting aside the planets Jupiter, Saturn, Uranus, and Neptune, each made entirely of gas and having no rigid surface, only four objects in our solar system have an atmosphere of any significance: Venus, Earth, Mars, and Titan.

Titan was not an accidental target of exploration. Its impressive résumé of molecules includes water, ammonia, methane, and ethane, as well as the multiringed compounds known as polycyclic aromatic hydrocarbons. The water ice is so cold it's as hard as concrete. But the combination of temperature and air pressure has liquefied the methane, and the first images sent back from *Huygens* seem to show streams, rivers, and lakes of the stuff. In some ways Titan's surface chemistry resembles that of the young Earth, which accounts for why so many astrobiologists view Titan as a "living" laboratory for studying Earth's distant past. Indeed, experiments conducted two decades ago show that adding water and a bit of acid to the organic ooze produced by irradiating the gases that make up Titan's hazy atmosphere yields sixteen amino acids.

Recently, biologists have learned that planet Earth may harbor a greater biomass belowground than on its surface. Ongoing investigations about the hardy habits of life demonstrate time and again that it recognizes few boundaries. Once stereotyped as kooky scientists in search of little green men on nearby planets, investigators who ponder the limits of life are now sophisticated hybrids, exploiting the tools of not only astrophysics, biology, and chemistry but also geology and paleontology as they pursue life here, there, and everywhere.

TWENTY-SIX

LIFE IN THE UNIVERSE

The discovery of hundreds of planets around stars other than the Sun has triggered tremendous public interest. Attention was driven not so much by the discovery of exosolar planets, but by the prospect of them hosting intelligent life. In any case, the media frenzy that continues may be somewhat out of proportion with the events. Why? Because planets cannot be all that rare in the universe if the Sun, an ordinary star, has at least eight of them. Also, the newly discovered planets are all oversized gaseous giants that resemble Jupiter, which means no convenient surface exists upon which life as we know it could live. And even if they were teeming with buoyant aliens, the odds against these life-forms being intelligent may be astronomical.

Ordinarily, there is no riskier step that a scientist (or anyone) can take than to make sweeping generalizations from just one example. At the moment, life on Earth is the only known life in the universe, but compelling arguments suggest we are not alone. Indeed, most astrophysicists accept the probability of life elsewhere. The reasoning is easy: if our solar system is not unusual, then there are so many planets in the universe that, for example, they outnumber the sum of all sounds and words ever uttered by every human who has ever lived. To declare that Earth must be the only planet with life in the universe would be inexcusably bigheaded of us.

Many generations of thinkers, both religious and scientific, have been led astray by anthropocentric assumptions, while others were simply led astray by ignorance. In the absence of dogma and data, it is safer to be guided by the notion that we are not special, which is generally known as the Copernican principle, named for Nicolaus Copernicus, of course, who, in the mid-1500s, put the Sun back in the middle of our solar system where it belongs. In spite of a third-century B.C. account of a Sun-centered universe, proposed by the Greek philosopher Aristarchus, the Earth-centered universe was by far the most popular view for most of the last 2,000 years. Codified by the teachings of Aristotle and Ptolemy, and later by the preachings of the Roman Catholic Church, people generally accepted Earth as the center of all motion and of the known universe. This fact was self-evident. The universe not only looked that way, but God surely made it so.

While the Copernican principle comes with no guarantees that it will forever guide us to cosmic truths, it's worked quite well so far: not only is Earth not in the center of the solar system, but the solar system is not in the center of the Milky Way galaxy, the Milky Way galaxy is not in the center of the universe, and it may come to pass that our universe is just one of many that comprise a multiverse. And in case you're one of those people who thinks that the edge may be a special place, we are not at the edge of anything either.

A WISE CONTEMPORARY posture would be to assume that life on Earth is not immune to the Copernican principle. To do so allows us ask how the appearance or the chemistry of life on Earth can provide clues to what life might be like elsewhere in the universe.

I do not know whether biologists walk around every day awestruck by the diversity of life. I certainly do. On this single planet called Earth, there coexist (among countless other life-forms), algae, beetles, sponges, jellyfish, snakes, condors, and giant sequoias. Imagine these seven living organisms lined up next to each other in size-place. If you didn't know better, you would be challenged to believe that they all came from the same universe, much less the same planet. Try describing a snake to somebody who has never seen one: "You gotta believe me. Earth has an animal that (1) can stalk its prey with infrared detectors, (2) swallows whole live animals up to five times bigger than its head, (3) has no arms, legs, or any other appendage, yet (4) can slide along level ground at a speed of two feet per second!"

Given this diversity of life on Earth, one might expect a diversity of life exhibited among Hollywood aliens. But I am consistently amazed by the film industry's lack of creativity. With a few notable exceptions such as the aliens of *The Blob* (1958), in *2001: A Space Odyssey* (1968), and in *Contact* (1997), Hollywood aliens look remarkably humanoid. No matter how ugly (or cute) they are, nearly all of them have two eyes, a nose, a mouth, two ears, a head, a neck, shoulders, arms, hands, fingers, a torso, two legs, two feet— and they can walk. From an anatomical view, these creatures are practically indistinguishable from humans, yet they are supposed to have come from another planet. If anything is certain, it is that life elsewhere in the universe, intelligent or otherwise, should look at least as exotic to us as some of Earth's own life-forms.

The chemical composition of Earth-based life is primarily derived from a select few ingredients. The elements hydrogen, oxygen, and carbon account for over 95 percent of the atoms in the human body and all known life. Of the three, the chemical structure of carbon allows it to bond readily and strongly with itself and with many other elements in many different ways, which is why we are considered to be carbon-based life, and which is why the study of molecules that contain carbon is generally known as "organic" chemistry. Curiously, the study of life elsewhere in the universe is known as exobiology, which is one of the few disciplines that attempts to function with the complete absence of firsthand data.

Is life chemically special? The Copernican principle suggests that it probably isn't. Aliens need not look like us to resemble us in more fundamental ways. Consider that the four most common elements in the universe are hydrogen, helium, carbon, and oxygen. Helium is inert. So the three most abundant, chemically active ingredients in the cosmos are also the top three ingredients in life on Earth. For this reason, you can bet that if life is found on another planet, it will be made of a similar mix of elements. Conversely, if life on Earth were composed primarily of, for example, molybdenum, bismuth, and plutonium, then we would have excellent reason to suspect that we were something special in the universe.

Appealing once again to the Copernican principle, we can assume that the size of an alien organism is not likely to be ridiculously large compared with life as we know it. There are cogent structural reasons why you would not expect to find a life the size of the Empire State Building strutting around a planet. But if we ignore these engineering limitations of biological matter we approach another, more fundamental limit. If we assume that an alien has control of its own appendages or, more generally, if we assume the organism functions coherently as a system, then its size would ultimately be constrained by its ability to send signals within itself at the speed of light—the fastest allowable speed in the universe. For an admittedly extreme example, if an organism were as big as the entire solar system (about 10 light-hours across), and if it wanted to scratch its head, then this simple act would take no less than 10 hours to accomplish. Sub-slothlike behavior such as this would be evolutionarily self-limiting because the time since the beginning of the universe may be insufficient for the creature to have evolved from smaller forms of life over many generations.

HOW ABOUT INTELLIGENCE? When Hollywood aliens manage to visit Earth, one might expect them to be remarkably smart. But I know of some that should have been embarrassed at their stupidity. During a four-hour car trip from Boston to New York City, while I was surfing the FM dial, I came upon a radio play in progress that, as best as I could determine, was about evil aliens who were terrorizing Earthlings. Apparently, they needed hydrogen atoms to survive so they kept swooping down to Earth to suck up its oceans and extract the hydrogen from all the H_2O molecules.

Now those were some dumb aliens.

They must not have been looking at other planets en route to Earth because Jupiter, for example, contains over two hundred times the entire mass of Earth in pure hydrogen. And I guess nobody ever told them that over 90 percent of all atoms in the universe are hydrogen.

And how about all those aliens that manage to traverse thousands of light-years through interstellar space, yet bungle their arrival by crash-landing on Earth?

Then there were the aliens in the 1977 film *Close Encounters of the Third Kind*, who, in advance of their arrival, beamed to Earth a mysterious sequence of repeated digits that encryption experts eventually decoded to be the latitude and longitude of the aliens' upcoming landing site. But Earth longitude has a completely arbitrary starting point—the prime meridian—which passes through Greenwich, England, by international agreement. And both longitude and latitude are measured in peculiar unnatural units we call degrees, 360 of which are in a circle. Armed with this much knowledge of human culture, it seems to me that the aliens could have just learned English and beamed the message, "We're going to land a little bit to the side of Devil's Tower National Monument in Wyoming. And since we're coming in a flying saucer we won't need the runway lights."

The award for dumbest creature of all time must go to the alien from the original 1983 film *Star Trek, The Motion Picture*. V-ger, as it called itself (pronounced vee-jer) was an ancient mechanical space probe that was on a mission to explore and discover and report back its findings. The probe was "rescued" from the depths of space by a civilization of mechanical aliens and reconfigured so that it could actually accomplish this mission for the entire universe. Eventually, the probe did acquire all knowledge and, in so doing, achieved consciousness. The *Enterprise* stumbles upon this now-sprawling monstrous collection of cosmic information at a time when the alien was searching for its original creator and the meaning of life. The stenciled letters on the side of the original probe revealed the characters *V* and *ger*. Shortly thereafter, Captain Kirk discovers that the probe was *Voyager 6*, which had been launched by humans on Earth in the late twentieth century. Apparently, the *oya* that fits between the *V* and the *ger* had been badly tarnished and was unreadable. Okay. But I have always wondered how *V-ger* could have acquired all knowledge of the universe *and* achieved consciousness yet not have known that its real name was *Voyager*.

And don't get me started on the 1996 summer blockbuster *Independence Day*. I find nothing particularly offensive about evil aliens. There would be no science-fiction film industry without them. The aliens in *Independence Day* were definitely evil. They looked like a genetic cross between a Portuguese Man of War jellyfish, a hammerhead shark, and a human being. While more creatively conceived than most Hollywood aliens, their flying saucers were equipped with upholstered high-back chairs and arm rests.

I'm glad that, in the end, the humans win. We conquer the Independence Day aliens by having a Macintosh laptop computer upload a software virus to the mothership (which happens to be one-fifth the mass of the Moon) to disarm its protective force field. I don't know about you, but I have trouble just uploading files to other computers within my own department, especially when the operating systems are different. There is only one solution. The entire defense system for the alien mothership must have been powered by the same release of Apple Computer's system software as the laptop computer that delivered the virus.

Thank you for indulging me. I had to get all that off my chest.

LET US ASSUME, for the sake of argument, that humans are the only species in the history of life on Earth to evolve high-level intelligence. (I mean no disrespect to other big-brained mammals.

While most of them cannot do astrophysics, or write poetry, my conclusions are not substantially altered if you wish to include them.) If life on Earth offers any measure of life elsewhere in the universe, then intelligence must be rare. By some estimates, there have been more than 10 billion species in the history of life on Earth. It follows that among all extraterrestrial life-forms we might expect no better than about 1 in 10 billion to be as intelligent as we are, not to mention the odds against the intelligent life having an advanced technology *and* a desire to communicate through the vast distances of interstellar space.

On the chance that such a civilization exists, radio waves would be the communication band of choice because of their ability to traverse the galaxy unimpeded by interstellar gas and dust clouds. But humans on Earth have only understood the electromagnetic spectrum for less than a century. More depressingly put, for most of human history, had aliens tried to send radio signals to Earthlings we would have been incapable of receiving them. For all we know, the aliens have already done this and unwittingly concluded that there was no intelligent life on Earth. They would now be looking elsewhere. A more humbling possibility would be if aliens had become aware of the technologically proficient species that now inhabits Earth, yet they had drawn the same conclusion.

Our life-on-Earth bias, intelligent or otherwise, requires us to hold the existence of liquid water as a prerequisite to life elsewhere. As already discussed, a planet's orbit should not be too close to its host star, otherwise the temperature would be too high and the planet's water content would vaporize. The orbit should not be too far away either, or else the temperature would be too low and the planet's water content would freeze. In other words, conditions on the planet must allow the temperature to stay within the 180 degree (Fahrenheit) range of liquid water. As in the three-bowls-of-food scene in the fairy tale *Goldilocks and the Three Bears,* the temperature has to be just right. When I was interviewed about this subject recently on a syndicated radio talk show, the host commented, "Clearly, what you should be looking for is a planet made of porridge!"

While distance from the host star is an important factor for the existence of life as we know it, other factors matter too, such as a planet's ability to trap stellar radiation. Venus is a textbook example of this "greenhouse" phenomenon. Visible sunlight that manages to pass through its thick atmosphere of carbon dioxide gets absorbed by Venus's surface and then reradiated in the infrared part of the spectrum. The infrared, in turn, gets trapped by the atmosphere. The unpleasant consequence is an air temperature that hovers at about 900 degrees Fahrenheit, which is much hotter than we would expect knowing Venus's distance to the Sun. At this temperature, lead swiftly liquefies.

The discovery of simple, unintelligent life-forms elsewhere in the universe (or evidence that they once existed) would be far more likely and, for me, only slightly less exciting than the discovery of intelligent life. Two excellent nearby places to look are the dried riverbeds of Mars, were there may be fossil evidence of life from when waters once flowed, and the subsurface oceans that are theorized to exist under the frozen ice layers of Jupiter's moon Europa. Once again, the promise of liquid water defines our targets of search.

Other commonly invoked prerequisites for the evolution of life in the universe involve a planet in a stable, nearly circular orbit around a single star. With binary and multiple star systems, which comprise about half of all "stars" in the galaxy, planet orbits tend to be strongly elongated and chaotic, which induces extreme temperature swings that would undermine the evolution of stable life-forms. We also require that there be sufficient time for evolution to run its course. High-mass stars are so short-lived (a few million years) that life on an Earthlike planet in orbit around them would never have a chance to evolve.

As we have already seen, the set of conditions to support life as we know it is loosely quantified through what is known as the Drake equation, named for the American astronomer Frank Drake. The Drake equation is more accurately viewed as a fertile idea than as a rigorous statement of how the physical universe works. It separates the overall probability of finding life in the galaxy into a set of simpler probabilities that correspond to our preconceived notions of the cosmic conditions that are suitable for life. In the end, after you argue with your colleagues about the value of each probability term in the equation, you are left with an estimate for the total number of intelligent, technologically proficient civilizations in the galaxy. Depending on your bias level, and your knowledge of biology, chemistry, celestial mechanics, and astrophysics, you may use it to estimate from at least one (we humans) up to millions of civilizations in the Milky Way.

IF WE CONSIDER the possibility that we may rank as primitive among the universe's technologically competent life-forms— however rare they may be—then the best we can do is keep alert for signals sent by others because it is far more expensive to send than to receive them. Presumably, an advanced civilization would have easy access to an abundant source of energy such as its host star. These are the civilizations that would be more likely to send rather than to receive. The search for extraterrestrial intelligence (affectionately known by its acronym "SETI") has taken many forms. The most advanced efforts today use a cleverly designed electronic detector that monitors, in its latest version, billions of radio channels in search of a signal that might rise above the cosmic noise.

The discovery of extraterrestrial intelligence, if and when it happens, will impart a change in human self-perception that may be impossible to anticipate. My only hope is that every other civilization isn't doing exactly what we are doing because then everybody would be listening, nobody would be receiving, and we would collectively conclude that there is no other intelligent life in the universe.

TWENTY-SEVEN

OUR RADIO BUBBLE

For the opening scene to the 1997 film *Contact*, a virtual camera executes a controlled, three-minute zoom from Earth to the outer reaches of the universe. For this journey, you happen to be equipped with receivers that enable you to decode Earth-based television and radio broadcasts that have escaped into space. Initially, you hear a cacophonous mixture of loud rock music, news broadcasts, and noisy static as though you were listening to dozens of radio stations simultaneously. As the journey progresses out into space, and as you overtake earlier broadcasts that have traveled farther, the signals become less cacophonous and distinctly older as they report historical events that span the broadcast era of modern civilization. Amid the noise, you hear sound bytes in reverse sequence that include: the *Challenger* shuttle disaster of January 1986; the Moon landing of July 20, 1969; Martin Luther King's famous "I Have a Dream" speech, delivered in August 28, 1963; President Kennedy's January 20, 1961, inaugural address; President Roosevelt's December 8, 1941, address to Congress, where he asked for a declaration of war; and a 1936 address by Adolf Hitler during his rise to power in Nazi Germany. Eventually, the human contribution to the signal disappears entirely, leaving a din of radio noise emanating from the cosmos itself.

Poignant. But this scroll of acoustic landmarks would not unfurl exactly as shown. If you somehow managed to violate several laws of physics and travel fast enough to overtake a radio wave, then few words would be intelligible because you'd hear everything replayed backward. Furthermore, we hear King's famous speech as we pass the planet Jupiter, implying Jupiter is as far as the broadcast has traveled. In fact, King's speech passed Jupiter 39 minutes after he delivered it.

Ignoring these facts that would render the zoom impossible, *Contact*'s opening scene was poetic and powerful, as it indelibly marked the extent to which we have presented our civilized selves to the rest of the Milky Way galaxy. This radio bubble, as it has come to be called, centers on Earth and continues to expand at the speed of light in every direction, while getting its center continuously refilled by modern broadcasts. Our bubble now extends nearly 100 light-years into space, with a leading edge that corresponds to the first artificial radio signals ever generated by Earthlings. The bubble's volume now contains about a thousand stars, including Alpha Centauri (4.3 light-years away), the nearest star system to the Sun; Sirius (10 light-years away), the brightest star in the nighttime sky; and every star around which a planet has thus far been discovered.

NOT ALL RADIO signals escape our atmosphere. The plasma properties of the ionosphere, more than 50 miles up, enable it to reflect back to Earth all radio-wave frequencies less than 20 megahertz, allowing some forms of radio communication, such as the well-known "short wave" frequencies of HAM radio operators, to reach thousands of miles beyond your horizon. All the broadcast frequencies of AM radio are also reflected back to Earth, accounting for the extended range that these stations enjoy.

If you broadcast at a frequency that does not correspond to those reflected by Earth's ionosphere, or if Earth had no ionosphere, your radio signals would reach only those receivers that fell in its line of "sight." Tall buildings give significant advantage to radio transmitters mounted on their roofs. While the horizon for a 5'8" person is just 3 miles away, the horizon seen by King Kong, while climbing atop New York City's Empire State Building, is more than 50. After the filming of that 1933 classic, a broadcast antenna was installed. An equally tall receiving antenna could, in principle, be located 50 miles farther still, enabling the signal to cross their mutual 50-mile horizon, thereby extending the signal's reach to 100 miles.

The ionosphere reflects neither FM radio nor broadcast television, itself a subset of the radio spectrum. As prescribed, they each travel no farther on Earth than the farthest receiver they can see, which allows cities that are relatively near each other to broadcast their own television programs. For this reason, television's local broadcasts and FM radio cannot possibly be as influential as AM radio, which may account for its preponderance of politically acerbic talk shows. But the real influence of FM and TV may not be terrestrial. While most of the signal's strength is purposefully broadcast horizontal to the ground, some of it leaks straight up, crossing the ionosphere and traveling through the depths of space. For them, the sky is not the limit. And unlike some other bands in the electromagnetic spectrum, radio waves have excellent penetration through the gas and dust clouds of interstellar space, so the stars are not the limit either.

If you add up all factors that contribute to the strength of Earth's radio signature, such as the total number of stations, the distribution of stations across Earth's surface, the energy output of each station, and the bandwidth over which the energy is broadcast, you find that television accounts for the largest sustained flux of radio signals detectable from Earth. The anatomy of a broadcast signal displays a skinny and a wide part. The skinny, narrow-band part is the video carrier signal, through which more than half the total energy is broadcast. At a mere .10 hertz wide in frequency, it establishes the station's location on the dial (the familiar channels 2 through 13) as well as the existence of the signal in the first place. A low-intensity, broadband signal, 5 million hertz wide, surrounds the carrier at higher and lower frequencies and is imbued with modulations that contain all the program information.

AS YOU MIGHT guess, the United States is the most significant contributor among all nations to Earth's global television profile. An eavesdropping alien civilization would first detect our strong carrier signals. If it continued to pay attention, it would notice periodic Doppler shifts in these signals (alternating from lower frequency to higher frequency) every 24 hours. It would then notice the signal get stronger and weaker over the same time interval. The aliens might first conclude that a mysterious, although naturally occurring, radio loud spot was rotating into and out of view. But if the aliens managed to decode the modulations within the surrounding broadband signal they would gain immediate access to elements of our culture.

Electromagnetic waves, including visible-light as well as radio waves, do not require a medium though which to travel. Indeed, they are happiest moving through the vacuum of space. So the time-honored flashing red sign in broadcast studios that says "On the Air" could justifiably read "Through Space," a phrase that applies especially to the escaping TV and FM frequencies.

As the signals move out into space they get weaker and weaker, becoming diluted by the growing volume of space through which it travels. Eventually, the signals get hopelessly buried by the ambient radio noise of the universe, generated by radio-emitting galaxies, the microwave background, radio-rich regions of star formation in the Milky Way, and cosmic rays. These factors, above all, will limit the likelihood of a distant civilization decoding our way of life.

At current broadcast strengths from Earth, aliens 100 light-years away would require a radio receiver that was fifteen times the collecting area of the 300-foot Arecibo telescope (the world's largest) to detect a television station's carrier signal. If they want to decode our programming information and hence our culture, they will need to compensate for the Doppler shifts caused by Earth's rotation on its axis and by its revolution around the Sun (enabling them to lock onto a particular TV station) and they must increase their detection capacity by another factor of 10,000 above that which would detect the carrier signal. In radio telescope terms, this amounts to a dish about four hundred times Arecibo's diameter, or about 20 miles across.

If technologically proficient aliens are indeed intercepting our signals (with a suitably large and sensitive telescope), and if they are managing to decode the modulations, then the basics of our culture would surely befuddle alien anthropologists. As they watch us become a radio-transmitting planet, their attention might first be flagged by early episodes of the *Howdy Doody* show. Once they knew to listen, they would then learn how typical human males and females interact with each other from episodes of Jackie Gleason's *Honeymooners* and from Lucy and Ricky in *I Love Lucy*. They might then assess our intelligence from episodes of *Gomer Pyle*, *The Beverly Hillbillies*, and then, perhaps, from *Hee Haw*. If the aliens didn't just give up at this point, and if they chose to wait a few more years, they would learn a little more about human interactions from Archie Bunker in *All in the Family*, then from George Jefferson in *The Jeffersons*. After a few more years of study, their knowledge would be further enriched from the odd characters in *Seinfeld* and, of course, the prime-time cartoon *The Simpsons*. (They would be spared the wisdom of the hit show *Beavis and Butthead* because it existed only as a nonbroadcast cable program on MTV.) These were among the most popular shows of our times, each sustaining cross-generational exposure in the form of reruns.

Mixed in among our cherished sitcoms is the extensive, decade-long news footage of bloodshed during the Vietnam war, the Gulf wars, and other military hot spots around the planet. After 50 years of television, there's no other conclusion the aliens could draw, but that most humans are neurotic, death-hungry, dysfunctional idiots.

IN THIS ERA of cable television, even broadcast signals that might have otherwise escaped the atmosphere are now delivered via wires directly to your home. There may come a time when television is no longer a broadcast medium, leaving our tube-watching aliens to wonder whether our species went extinct.

For better or for worse, television might not be the only signals from Earth decoded by aliens. Any time we communicate with our astronauts or our space probes, all signals that do not intersect the craft's receiver are lost in space forever. The efficiency of this communication is greatly improved by modern methods of signal compression. In the digital era, it's all about bytes per second. If you devised a clever algorithm that compressed your signal by a factor of 10, you could communicate ten times more efficiently, provided the person or machine on the other side of the signal knew how to undo your compressed signal. Modern examples of compression utilities include those that create MP acoustic recordings, JPEG images, and MPEG movies for your computer, enabling you to swiftly transfer files and to reduce the clutter on your hard drive.

The only radio signal that cannot be compressed is one that contains completely random information, leaving it indistinguishable from radio static. In a related fact, the more you compress a signal, the more random it looks to someone who intercepts it. A perfectly compressed signal will, in fact, be indistinguishable from static to everyone but the person who has the preordained knowledge and resources to decode it. What does it all mean? If a culture is sufficiently advanced and efficient, then their signals (even without the influence of cable transmissions) might just disappear completely from the cosmic highways of gossip.

Ever since the invention and widespread use of electric bulbs, human culture has also created a bubble in the form of visible light. This, our nighttime signature, has slowly changed from tungsten incandescence to neon from billboards and sodium from the now-widespread use of sodium vapor lamps for streetlights. But apart from the Morse code flashed by shuttered lamps from the decks of ships, we typically do not send visible light through the air to carry signals, so our visual bubble is not interesting. It's also hopelessly lost in the visible-light glare of our Sun.

RATHER THAN LET aliens listen to our embarrassing TV shows, why not send them a signal of our own choosing, demonstrating how intelligent and peace loving we are? This was first done in the form of gold-etched plaques affixed to the sides of the four unmanned planetary probes *Pioneer 10* and *11* and *Voyager 1* and *2*. Each plaque contains pictograms conveying our base of scientific knowledge and our location in the Milky Way galaxy while the *Voyager* plaques also contain audio information about the kindness of our species. At 50,000 miles per hour—a speed in excess of the solar system's escape velocity—these spacecraft are traveling through interplanetary space at quite a clip. But they move ridiculously slow compared with the speed of light and won't get to the nearby stars for another 100,000 years. They represent our "spacecraft" bubble. Don't wait up for them.

A better way to communicate is to send a high-intensity radio signal to a busy place in the galaxy, like a star cluster. This was first done in 1976, when the Arecibo telescope was used in reverse, as a transmitter rather than a receiver, to send the first radio-wave signal of our own choosing out to space. That message, at the time of this writing, is now 30 light-years from Earth, headed in the direction of the spectacular globular star cluster known as M13, in the constellation Hercules. The message contains in digital form some of what appeared on the *Pioneer* and *Voyager* spacecraft. Two problems, however: The globular cluster is so chock full of stars (at least a half-million) and so tightly packed, that planetary orbits tend to be unstable as their gravitational allegiance to their host star is challenged for every pass through the cluster's center. Furthermore, the cluster has such a meager quantity of heavy elements (out of which planets are made) that planets are probably rare in the first place.

These scientific points were not well known or understood at the time the signal was sent.

In any case, the leading edge of our "on-purpose" radio signals (forming a directed radio cone, instead of a bubble) is 30 light-years away and, if intercepted, may mend the aliens' image of us based on the radio bubble of our television shows. But this will happen only if the aliens can somehow determine which type of signal comes closer to the truth of who we are, and what our cosmic identity deserves to be.

WHEN THE UNIVERSE TURNS BAD

ALL THE WAYS THE COSMOS WANTS TO KILL US

TWENTY-EIGHT

CHAOS IN THE SOLAR SYSTEM

S cience distinguishes itself from almost all other human endeavors by its capacity to predict future events with precision. Daily newspapers often give you the dates for upcoming phases of the moon or the time of tomorrow's sunrise. But they do not tend to report "news items of the future" such as next Monday's closing prices on the New York Stock Exchange or next Tuesday's plane crash. The general public knows intuitively, if not explicitly, that science makes predictions, but it may surprise people to learn that science can also predict that something is unpredictable. Such is the basis of chaos. And such is the future evolution of the solar system.

A chaotic solar system would, no doubt, have upset the German astronomer Johannes Kepler, who is generally credited with the first predictive laws of physics, published in 1609 and 1619. Using a formula that he derived empirically from planetary positions on the sky, he could predict the average distance between any planet and the Sun by simply knowing the duration of the planet's year. In Isaac Newton's 1687 *Principia*, his universal law of gravity allows you to mathematically derive all of Kepler's laws from scratch.

In spite of the immediate success of his new laws of gravity, Isaac Newton remained concerned that the solar system might one day fall into disarray. With characteristic prescience, Newton noted in Book III of his 1730 edition of *Optiks*:

> The Planets move one and the same way in Orbs concentric, some
> inconsiderable Irregularities excepted, which may have arisen from
> the mutual actions of . . . Planets upon one another, and which will
> be apt to increase, till the system wants a Reformation. (p. 402)

As we will detail in Section 7, Newton implied that God might occasionally be needed to step in and fix things. The celebrated French mathematician and dynamicist Pierre-Simon Laplace had an opposite view of the world. In his 1799–1825 five-volume treatise *Traité de mécanique céleste*, he was convinced that the universe was stable and fully predictable. Laplace later wrote in *Philosophical Essays on Probability* (1814):

> [With] all the forces by which nature is animated . . . nothing [is]
> uncertain, and the future as the past would be present to [one's] eyes.
> (1995, Chap. II, p. 3)

The solar system does, indeed, look stable if all you have at your disposal is a pencil and paper. But in the age of supercomputers, where billions of computations per second are routine, solar system models can be followed for hundreds of millions of years. What thanks do we get for our deep understanding of the universe?

Chaos.

Chaos reveals itself through the application of our well-tested physical laws in computer models of the solar system's future evolution. But has also reared its head in other disciplines, such as meteorology and predator-prey ecology, and almost anyplace where you find complex interacting systems.

To understand chaos as it applies to the solar system, one must first recognize that the difference in location between two objects, commonly known as their distance, is just one of many differences that can be calculated. Two objects can also differ in energy, orbit size, orbit shape, and orbit inclination. One could therefore broaden the concept of distance to include the separation of objects in these other variables as well. For example, two objects that are (at the moment) near each other in space may have very different orbit shapes. Our modified measure of "distance" would tell us that the two objects are widely separated.

A common test for chaos is to begin with two computer models that are identical in every way except for a small change somewhere. In one of two solar system models you might allow Earth to recoil slightly in its orbit from being hit by a small meteor. We are now armed to ask a simple question: Over time, what happens to the "distance" between these two nearly identical models? The distance may remain stable, fluctuate, or even diverge. When two models diverge exponentially, they do so because the small differences between them magnify over time, badly confounding your ability to predict the future. In some cases, an object can be ejected from the solar system completely.

This is the hallmark of chaos.

For all practical purposes, in the presence of chaos, it is *impossible* to reliably predict the distant future of the system's evolution. We owe much of our early understanding of the onset of chaos to Alexander Mikhailovich Lyapunov (1857–1918), who was a Russian mathematician and mechanical engineer. His 1892 PhD thesis "The General Problem of the Stability of Motion" remains a classic to this day. (By the way, Lyapunov died a violent death in the chaos of political unrest that immediately followed the Russian Revolution.)

Since the time of Newton, people knew that you can calculate the exact paths of two isolated objects in mutual orbit, such as a binary star system, for all of time. No instabilities there. But as you add more objects to the dance card, orbits become more and more complex, and more and more sensitive to their initial conditions. In the solar system we have the Sun, its nine planets, their 70+ satellites, asteroids, and comets. This may sound complicated enough, but the story is not yet complete. Orbits in the solar system are further influenced by the Sun's loss of 4 million tons of matter every second from the thermonuclear fusion in its core. The matter converts to energy, which subsequently escapes as light from the Sun's surface. The Sun also loses mass from the continuously ejected stream of charged particles known as the solar wind. And the solar system is further subject to the perturbing gravity from stars that occasionally pass by in their normal orbit around the galactic center.

To appreciate the task of the solar system dynamicist, consider that the equations of motion allow you to calculate the net force of gravity on an object, at any given instant, from all other known objects in the solar system and beyond. Once you know the force on each object, you nudge them all (on the computer) in the direction they ought to go. But the force on each object in the solar system is now slightly different because everybody has moved. You must therefore recompute all forces and nudge them again. This continues for the duration of the simulation, which in some cases involves trillions of nudges. When you do these calculations, or ones similar to them, the solar system's behavior is chaotic. Over time intervals of about 5 million years for the inner terrestrial planets (Mercury, Venus, Earth, and Mars) and about 20 million years for the outer gas giants (Jupiter, Saturn, Uranus, and Neptune), arbitrarily small "distances" between initial conditions noticeably diverge. By 100 to 200 million years into the model, we have lost all ability to predict planet trajectories.

Yes, this is bad. Consider the following example: The recoil of Earth from the launch of a single space probe can influence our future in such a way that in about 200 million years, the position of Earth in its orbit around the Sun will be shifted by nearly 60 degrees. For the distant future, surely it's just benign ignorance if we do not know where Earth will be in its orbit. But tension rises when we realize that asteroids in one family of orbits can chaotically migrate to another family of orbits. If asteroids can migrate, and if Earth can be somewhere in its orbit that we cannot predict, then there is a limit to how far in the future we can reliably calculate the risk of a major asteroid impact and the global extinction that might ensue.

Should the probes we launch be made of lighter materials? Should we abandon the space program? Should we worry about solar mass loss? Should we be concerned about the thousand tons of meteor dust per day that Earth accumulates as it plows through the debris of interplanetary space? Should we all gather on one side of Earth and leap into space together? None of the above. The long-term effects of these small variations are lost in the chaos that unfolds. In a few cases, ignorance in the face of chaos can work to our advantage.

A skeptic might worry that the unpredictability of a complex, dynamic system over long time intervals is due to a computational round-off error, or some peculiar feature of the computer chip or computer program. But if this suspicion were true, then two-object systems might eventually show chaos in the computer models. But they don't. And if you pluck Uranus from the solar system model and repeat the orbit calculations for the gas giant planets, then there is no chaos. Another test comes from computer simulations of Pluto, which has a high eccentricity and an embarrassing tilt to its orbit. Pluto actually exhibits well-behaved chaos, where small "distances" between initial conditions lead to an unpredictable yet limited set of trajectories. Most importantly, however, different investigators using different computers and different computational methods have derived similar time intervals for the onset of chaos in the long-term evolution of the solar system.

Apart from our selfish desire to avoid extinction, broader reasons exist for studying the long-term behavior of the solar system. With a full evolutionary model, dynamicists can go backward in time to probe the history of the solar system, when the planetary roll call may have been very different from today. For example, some planets that existed at the birth of the solar system (5 billion years ago) could have since been forcibly ejected. Indeed we may have begun with several dozen planets, instead of eight, having lost most of them jack-in-the-box style to interplanetary space.

In the past four centuries, we have gone from not knowing the motions of the planets to knowing that we cannot know the evolution of the solar system into the unlimited future—a bittersweet victory in our unending quest to understand the universe.

TWENTY-NINE

COMING ATTRACTIONS

One needn't look far to find scary predictions of a global holocaust by killer asteroids. That's good, because most of what you might have seen, read, or heard is true.

The chances that your or my tombstone will read "killed by asteroid" are about the same for "killed in an airplane crash." About two dozen people have been killed by falling asteroids in the past 400 years, but thousands have died in crashes during the relatively brief history of passenger air travel. So how can this comparative statistic be true? Simple. The impact record shows that by the end of 10 million years, when the sum of all airplane crashes has killed a billion people (assuming a death-by-airplane rate of 100 per year), an asteroid is likely to have hit Earth with enough energy to kill a billion people. What confuses the interpretation is that while airplanes kill people a few at a time, our asteroid might not kill anybody for millions of years. But when it hits, it will take out hundreds of millions of people instantaneously and many more hundreds of millions in the wake of global climatic upheaval.

The combined asteroid and comet impact rate in the early solar system was frighteningly high. Theories and models of planet formation show that chemically rich gas condenses to form molecules, then particles of dust, then rocks and ice. Thereafter, it's a shooting gallery. Collisions serve as a means for chemical and gravitational forces to bind smaller objects into larger ones. Those objects that, by chance, accreted slightly more mass than average will have slightly higher gravity and attract other objects even more. As accretion continues, gravity eventually shapes blobs into spheres and planets are born. The most massive planets had sufficient gravity to retain gaseous envelopes. All planets continue to accrete for the rest of their days, although at a significantly lower rate than when formed.

Still, billions (likely trillions) of comets remain in the extreme outer solar system, up to a thousand times the size of Pluto's orbit, that are susceptible to gravitational nudges from passing stars and interstellar clouds that set them on their long journey inward toward the Sun. Solar system leftovers also include short-period comets, of which several dozen are known to cross Earth's orbit, and thousands of asteroids that do the same.

The term "accretion" is duller than "species-killing, ecosystem-destroying impact." But from the point of view of solar system history, the terms are the same. We cannot simultaneously be happy we live on a planet; happy that our planet is chemically rich; and happy we are not dinosaurs; yet resent the risk of planetwide catastrophe. Some of the energy from asteroid collisions with Earth gets dumped into our atmosphere through friction and an airburst of shock waves. Sonic booms are shock waves too, but they are typically made by airplanes with speeds anywhere between one and three times the speed of sound. The worst damage they might do is jiggle the dishes in your cabinet. But with speeds upwards of 45,000 miles per hour—nearly seventy times the speed of sound—the shock waves from your average collision between an asteroid and Earth can be devastating.

If the asteroid or comet is large enough to survive its own shock waves, the rest of its energy gets deposited on Earth's surface in an explosive event that melts the ground and blows a crater that can measure twenty times the diameter of the original object. If many impactors were to strike with little time between each event, then Earth's surface would not have enough time to cool between impacts. We infer from the pristine cratering record on the surface of the Moon (our nearest neighbor in space) that Earth experienced an era of heavy bombardment between 4.6 and 4 billion years ago. The oldest fossil evidence for life on Earth dates from about 3.8 billion years ago. Not much before that, Earth's surface was unrelentingly sterilized, and so the formation of complex molecules, and thus life, was inhibited. In spite of this bad news, all the basic ingredients were being delivered nonetheless.

How long did life take to emerge? An often-quoted figure is 800 million years (4.6 billion − 3.8 billion = 800 million). But to be fair to organic chemistry, you must first subtract all the time Earth's surface was forbiddingly hot. That leaves a mere 200 million years for life to emerge from a rich chemical soup, which, as do all good soups, includes water.

Yes, the water you drink each day was delivered to Earth in part by comets more than 4 billion years ago. But not all space debris is left over from the beginning of the solar system. Earth has been hit at least a dozen times by rocks ejected from Mars, and we've been hit countless more times by rocks ejected from the Moon. Ejection occurs when impactors carry so much energy that smaller rocks near the impact zone get thrust upward with sufficient speed to escape the gravitational grip of the planet. Afterward, the rocks mind their own ballistic business in orbit around the Sun until they slam into something else. The most famous of the Mars rocks is the first meteorite found near the Alan Hills section of Antarctica in 1984. Officially known by its coded, though sensible, abbreviation, ALH-84001, this meteorite contains tantalizing, though circumstantial, evidence that simple life on the Red Planet thrived a billion years ago. Mars bears boundless geological evidence for a history of running water that includes dried riverbeds, river deltas, and flood plains. And most recently the Martian rovers *Spirit* and *Opportunity* found rocks and minerals that could have formed only in the presence of standing water.

Since liquid water is crucial to the survival of life as we know it, the possibility of life on Mars does not stretch scientific credulity. The fun part comes when you speculate whether life arose on Mars first, was blasted off its surface as the solar system's first bacterial astronauts, and then arrived to jump-start Earth's own evolution of life. There's even a word for the process: panspermia. Maybe we are all descendants of Martians.

Matter is far more likely to travel from Mars to Earth than vice versa. Escaping Earth's gravity requires over two-and-a-half times the energy than that required to leave Mars. Furthermore, Earth's atmosphere is about a hundred times denser. Air resistance on Earth (relative to Mars) is formidable. In any case, bacteria would have to be hardy indeed to survive the several million years of inter-planetary wanderings before landing on Earth. Fortunately, there is no shortage of liquid water and rich chemistry on Earth, so we do not require theories of panspermia to explain the origin of life as we know it, even if we still cannot explain it.

Ironically, we can (and do) blame impacts for major episodes of extinction in the fossil record. But what are the current risks to life and society? Below is a table that relates average collision rates on Earth with the size of impactor and the equivalent energy in millions of tons of TNT. For reference, I include a column that compares the impact energy in units of the atomic bomb that the United States dropped on the city of Hiroshima in 1945. These data are adapted from a graph by NASA's David Morrison (1992).

Once per ...	Asteroid Diameter (meters)	Impact Energy (Megatons of TNT)	Impact Energy (A-Bombs)
Month	3	0.001	0.05
Year	6	0.01	0.5
Decade	15	0.2	10
Century	30	2	100
Millennium	100	50	2,500
10,000 years	200	1,000	50,000
1,000,000 years	2000	1,000,000	50,000,000
100,000,000 years	10,000	100,000,000	5,000,000,000

The table is based on a detailed analysis of the history of impact craters on Earth, the erosion-free cratering record on the Moon's surface, and the known numbers of asteroids and comets whose orbits cross that of Earth.

The energetics of some famous impacts can be located on the table. For example, a 1908 explosion near the Tunguska River, Siberia, felled thousands of square kilometers of trees and incinerated the 300 square kilometers that encircled ground zero. The impactor is believed to have been a 60-meter stony meteorite (about the size of a 20-story building) that exploded in midair, thus leaving no crater. The chart predicts collisions of this magnitude to happen, on average, every couple of centuries. The 200-kilometer diameter Chicxulub Crater in the Yucatan, Mexico, is believed to be the calling card of a 10-kilometer asteroid. With an impact energy 5 billion times greater than the atomic bombs exploded in World War II, such a collision is predicted to occur about once in about 100 million years. The crater dates from 65 million years ago, and there hasn't been one of its magnitude since. Coincidentally, at about the same time, Tyrannosaurus rex and friends became extinct, enabling mammals to evolve into something more ambitious than tree shrews.

Those paleontologists and geologists who remain in denial of the role of cosmic impacts in the extinction record of Earth's species must figure out what else to do with the deposit of energy being delivered to Earth from space. The range of energies varies astronomically. In a review of the impact hazard to Earth written for the fat book *Hazards Due to Comets and Asteroids* (Gehrels 1994), David Morrison of NASA Ames Research Center, Clark R. Chapman of the Planetary Science Institute, and Paul Slovic of the University of Oregon briefly describe the consequence of unwelcome deposits of energy to Earth's ecosystem. I adapt what follows from their discussion.

Most impactors with less than about 10 megatons of energy will explode in the atmosphere and leave no trace of a crater. The few that survive in one piece are likely to be iron-based.

A 10- to 100-megaton blast from an iron asteroid will make a crater, while its stony equivalent will disintegrate and produce primarily air bursts. A land impact will destroy the area equivalent to that of Washington, DC.

Land impacts between 1,000 and 10,000 megatons continue to produce craters; oceanic impacts produce significant tidal waves. A land impact can destroy an area the size of Delaware.

A 100,000- to 1,000,000-megaton blast will result in global destruction of ozone; oceanic impacts will generate tidal waves felt on an entire hemisphere of Earth while land impacts raise enough dust into the stratosphere to change Earth's climate and freeze crops. A land impact will destroy an area the size of France.

A 10,000,000- to 100,000,000-megaton blast results in prolonged climactic effects and global conflagration. A land impact will destroy an area equivalent to the continental United States.

A land or ocean impact of 100,000,000 to 1,000,000,000 megatons will lead to mass extinction on a scale the Chicxulub impact 65 million years ago, when nearly 70 percent of Earth's species were suddenly wiped out.

Fortunately, among the population of Earth-crossing asteroids, we have a chance at cataloging everything larger than about a kilometer—the size that begins to wreak global catastrophe. An early-warning and defense system to protect the human species from these impactors is a realistic goal, as was recommended in NASA's *Spaceguard Survey Report*, and, believe it or not, continues to be on the radar screen of Congress. Unfortunately, objects smaller than about a kilometer do not reflect enough light to be reliably and thoroughly detected and tracked. These can hit us without notice, or they can hit with notice that is much too short for us to do anything about. The bright side of this news is that while they have enough energy to create local catastrophe by incinerating entire nations, they will not put the human species at risk of extinction.

Of course Earth is not the only rocky planet at risk of impacts. Mercury has a cratered face that, to a casual observer, looks just like the Moon. Recent radio topography of cloud-enshrouded Venus shows plenty of craters too. And Mars, with its historically active geology, reveals large craters that were recently formed.

At over three hundred times the mass of Earth, and at over ten times its diameter, Jupiter's ability to attract impactors is unmatched among the planets in the solar system. In 1994, during the week of anniversary celebrations for the 25th anniversary of the *Apollo 11* Moon landing, comet Shoemaker-Levy 9, having been broken apart into two dozen chunks during a previous close-encounter with Jupiter, slammed, one piece after another, into the Jovian atmosphere. The gaseous scars were seen easily from Earth with backyard telescopes. Because Jupiter rotates quickly (once every 10 hours), each part of the comet fell in a different location as the atmosphere rotated by.

And, in case you were wondering, each piece hit with the equivalent energy of the Chicxulub impact. So, whatever else is true about Jupiter, it surely has no dinosaurs left!

Earth's fossil record teems with extinct species—life-forms that had thrived far longer than the current Earth-tenure of *Homo sapiens*. Dinosaurs are in this list. What defense do we have against such formidable impact energies? The battle cry of those with no nuclear war to fight is "nuke them from the sky." True, the most efficient package of destructive energy ever conceived by humans is nuclear power. A direct hit on an incoming asteroid might explode it into enough small pieces to reduce the impact danger to a harmless, though spectacular, meteor shower. Note that in empty space, where there is no air, there can be no shock waves, so a nuclear warhead must actually make contact with the asteroid to do damage.

Another method is to engage those radiation-intensive neutron bombs (you remember—they were the bombs that killed people but left the buildings standing) in a way that the high-energy neutron bath heats one side of the asteroid to sufficient temperature that material spews forth and the asteroid recoils out of the collision path. A kindler, gentler method is to nudge the asteroid out of harm's way with slow but steady rockets that are somehow attached to one side. If you do this early enough, then all you need is a small push using conventional chemical fuels. If we catalogued every single kilometer-sized (and larger) object whose orbit intersects Earth's, then a detailed computer calculation would enable us to predict a catastrophic collision hundreds, and even thousands, of orbits in the future, granting Earthlings sufficient time to mount an appropriate defense. But our list of potential killer impactors is woefully incomplete, and chaos severely compromises our ability to predict the behavior of objects for millions and billions of orbits into the future.

In this game of gravity, by far the scariest breed of impactor is the long-period comet, which, by convention, are those with periods greater than 200 years. Representing about one-fourth of Earth's total risk of impacts, they fall toward the inner solar system from great distances and achieve speeds in excess of 100,000 miles per hour by the time they reach Earth. Long-period comets thus achieve more awesome impact energy for their size than your run-of-the-mill asteroid. More importantly, they are too dim over most of their orbit to be reliably tracked. By the time a long-period comet is discovered to be heading our way, we might have anywhere from several months to two years to fund, design, build, launch, and intercept it. For example, in 1996, comet Hyakutake was discovered only four months before its closest approach to the Sun because its orbit was tipped strongly out of the plane of our solar system, precisely where nobody was looking. While en route, it came within 10 million miles of Earth (a narrow miss) and made for spectacular nighttime viewing.

And here's one for your calendar: On Friday the 13th of April, 2029, an asteroid large enough to fill the Rose Bowl as though it were an egg cup, will fly so close to Earth that it will dip below the altitude of our communication satellites. We did not name this asteroid Bambi. Instead, it's named Apophis, after the Egyptian god of darkness and death. If the trajectory of Apophis at close approach passes within a narrow range of altitudes called the "keyhole," the precise influence of Earth's gravity on its orbit will guarantee that seven years later in 2036, on its next time around, the asteroid will hit Earth directly, slamming in the Pacific Ocean between California and Hawaii. The tsunami it creates will wipe out the entire west coast of North America, bury Hawaii, and devastate all the land masses of the Pacific Rim. If Apophis misses the keyhole in 2029, then, of course, we have nothing to worry about in 2036.

Should we build high-tech missiles that live in silos somewhere awaiting their call to defend the human species? We would first need that detailed inventory of the orbits of all objects that pose a risk to life on Earth. The number of people in the world engaged in this search totals a few dozen. How long into the future are you willing to protect Earth? If humans one day become extinct from a catastrophic collision, there would be no greater tragedy in the history of life in the universe. Not because we lacked the brain power to protect ourselves but because we lacked the foresight. The dominant species that replaces us in postapocalyptic Earth just might wonder, as they gaze upon our mounted skeletons in their natural history musems, why large-headed *Homo sapiens* fared no better than the proverbially pea-brained dinosaurs.

THIRTY

ENDS OF THE WORLD

Sometimes it seems that everybody is trying to tell you when and how the world will end. Some scenarios are more familiar than others. Those that are widely discussed in the media include rampant infectious disease, nuclear war, collisions with asteroids or comets, and environmental blight. While different from one another, they can all can bring about the end of the human species (and perhaps selected other life-forms) on Earth. Indeed, clichéd slogans such as "Save the Earth" contain the implicit call to save life on Earth, and not the planet itself. Fact is, humans cannot really kill Earth. Our planet will remain in orbit around the Sun, along with its planetary brethren, long after *Homo sapiens* has become extinct by whatever cause.

What hardly anybody talks about are end-of-world scenarios that do, in fact, jeopardize our temperate planet in its stable orbit around the Sun. I offer these prognostications not because humans are likely to live long enough to observe them, but because the tools of astrophysics enable me to calculate them. Three that come to mind are the death of the Sun, the impending collision between our Milky Way galaxy and the Andromeda galaxy, and the death of the universe, about which the community of astrophysicists has recently achieved consensus.

Computer models of stellar evolution are akin to actuarial tables. They indicate a healthy 10-billion-year life expectancy for our Sun.

At an estimated age of 5 billion years, the Sun will enjoy another 5 billion years of relatively stable energy output. By then, if we have not figured out a way to leave Earth, we will be around when the Sun exhausts its fuel supply. At that time, we will bear witness to a remarkable yet deadly episode in a star's life.

The Sun owes its stability to the controlled fusion of hydrogen into helium in its 15-million-degree core. The gravity that wants to collapse the star is held in balance by the outward gas pressure that the fusion sustains. While more than 90 percent of the Sun's atoms are hydrogen, the ones that matter reside in the Sun's core. When the core exhausts its hydrogen, all that will be left there is a ball of helium atoms that require an even higher temperature than does hydrogen to fuse into heavier elements. With its central engine temporarily shut off, the Sun will go out of balance. Gravity wins, the inner regions of the star collapse, and the central temperature rises through 100 million degrees, triggering the fusion of helium into carbon.

Along the way, the Sun's luminosity grows astronomically, which forces its outer layers to expand to bulbous proportions, engulfing the orbits of Mercury and Venus. Eventually, the Sun will swell to occupy the entire sky as its expansion subsumes the orbit of Earth. Earth's surface temperature will rise until it matches the 3,000-degree rarefied outer layers of the expanded Sun. Our oceans will come to a rolling boil as they evaporate entirely into interplanetary space. Meanwhile, our heated atmosphere will evaporate as Earth becomes a red-hot, charred ember orbiting deep within the Sun's gaseous outer layers. These layers will impede orbit, forcing Earth to trace a rapid death spiral down toward the Sun's core. As Earth descends, sinking nearer and nearer to the center, the Sun's rapidly rising temperature simply vaporizes all traces of our planet. Shortly thereafter, the Sun will cease all nuclear fusion; lose its tenuous, gaseous envelope containing Earth's scattered atoms; and expose its dead central core.

But not to worry. We will surely go extinct for some other reason long before this scenario unfolds.

NOT LONG AFTER the Sun terrorizes Earth, the Milky Way will encounter some problems of its own. Of the hundreds of thousands of galaxies whose velocity relative to the Milky Way has been reliably measured, only a few are moving toward us while all the rest are moving away at a speed directly related to their distances from us. Discovered in the 1920s by Edwin Hubble, after whom the *Hubble Space Telescope* was named, the general recession of galaxies is the observational signature of our expanding universe. The Milky Way and the several-hundred-billion-star Andromeda galaxy are close enough to each other that the expanding universe has a negligible effect on their relative motions. Andromeda and the Milky Way happen to be drifting toward each other at about 100 kilometers per second (a quarter-million miles per hour). If our (unknown) sideways motion is small, then at this rate, the 2.4-million light-year distance that separates us will shrink to zero within about 7 billion years.

Interstellar space is so vast and empty that there is no need to worry about stars in the Andromeda galaxy accidentally slamming into the Sun. During the galaxy-galaxy encounter, which would be a spectacular sight from a safe distance, stars are likely to pass each other by. But the event would not be worry-free. Some of Andromeda's stars could swing close enough to our solar system to influence the orbit of the planets and of the hundreds of billions of resident comets in the outer solar system. For example, close stellar flybys can throw one's gravitational allegiance into question. Computer simulations commonly show that the planets are either stolen by the interloper in a "flyby looting" or they become unbound and get flung into interplanetary space.

Back in Section 4, remember how choosy Goldilocks was with other people's porridge? If Earth gets stolen by the gravity of another star, there's no guarantee that our newfound orbit will be at the right distance to sustain liquid water on Earth's surface—a condition generally agreed to be a prerequisite to sustaining life as we know it. If Earth orbits too close, its water supply evaporates. And if Earth orbits too far, its water supply freezes solid.

If, by some miracle of future technology, Earth's inhabitants manage to prolong the Sun's life, then these efforts will be rendered irrelevant when Earth is flung into the cold depths of space. The absence of a nearby energy source will allow Earth's surface temperature to drop swiftly to hundreds of degrees below zero Fahrenheit. Our cherished atmosphere of nitrogen and oxygen and other gases would first liquefy and then drop to the surface and freeze solid, encrusting the Earth like icing on a spherical cake. We would freeze to death before we had a chance to starve to death. The last surviving life on Earth would be those privileged organisms that had evolved to rely not on the Sun's energy but on (what will then be) weak geothermal and geochemical sources, deep beneath the surface, in the cracks and fissures of Earth's crust. At the moment, humans are not among them.

One way to escape this fate is to fire up the warp drives and, like a hermit crab and snail shells, find another planet elsewhere in the galaxy to call home.

WITH OR WITHOUT warp drives, the long-term fate of the cosmos cannot be postponed or avoided. No matter where you hide, you will be part of a universe that inexorably marches toward a peculiar oblivion. The latest and best evidence available on the space density of matter and energy and the expansion rate of the universe suggest that we are on a one-way trip: the collective gravity of everything in the universe is insufficient to halt and reverse the cosmic expansion.

The most successful description of the universe and its origin combines the big bang with our modern understanding of gravity, derived from Einstein's general theory of relativity. As we will see in Section 7, the very early universe was a trillion-degree maelstrom of matter mixed with energy. During the 14-billion-year expansion that followed, the background temperature of the universe has dropped to a mere 2.7 degrees on the absolute (Kelvin) temperature scale. And as the universe continues to expand, this temperature will continue to approach zero.

Such a low background temperature does not directly affect us on Earth because our Sun (normally) grants us a cozy life. But as each generation of stars is born from clouds of interstellar gas, less and less gas remains to comprise the next generation of stars. This precious gas supply will eventually run out, as it already has in nearly half the galaxies in the universe. The small fraction of stars with the highest mass will collapse completely, never to be seen again. Some stars end their lives by blowing their guts across the galaxy in a supernova explosion. This returned gas can then be tapped for the next generation. But the majority of stars—Sun included—ultimately exhaust the fuel at their cores and, after the bulbous giant phase, collapse to form a compact orb of matter that radiates its feeble leftover heat to the frigid universe.

The short list of corpses may sound familiar: black holes, neutron stars (pulsars), and white dwarfs are each a dead end on the evolutionary tree of stars. But what they each have in common is an eternal lock on the material of cosmic construction. In other words, if stars burn out and no new ones are formed to replace them, then the universe will eventually contain no living stars.

How about Earth? We rely on the Sun for a daily infusion of energy to sustain life. If the Sun and the energy from all other stars were cut off from us then mechanical and chemical processes (life included) on and within Earth would "wind down." Eventually, the energy of all motion gets lost to friction and the system reaches a single uniform temperature. Earth, sitting beneath starless skies, will lie naked in the presence of the frozen background of the expanding universe. The temperature on Earth will drop, the way a freshly baked apple pie cools on a windowsill. Yet Earth is not alone in this fate. Trillions of years into the future, when all stars are gone, and every process in every nook and cranny of the expanding universe has wound down, all parts of the cosmos will cool to the same temperature as the ever-cooling background. At that time, space travel will no longer provide refuge because even Hell will have frozen over.

We may then declare that the universe has died—not with a bang, but with a whimper.

GALACTIC ENGINES

Galaxies are phenomenal objects in every way. They are the fundamental organization of visible matter in the universe. The universe contains as many as a hundred billion of them. They each commonly pack hundreds of billions of stars. They can be spiral, elliptical, or irregular in shape. Most are photogenic. Most fly solo in space, while others orbit in gravitationally linked pairs, familial groups, clusters, and superclusters.

The morphological diversity of galaxies has prompted all manner of classification schemes that supply a conversational vocabulary for astrophysicists. One variety, the "active" galaxy, emits an unusual amount of energy in one or more bands of light from the galaxy's center. The center is where you will find a galactic engine. The center is where you will find a supermassive black hole.

The zoo of active galaxies reads like the manifest for a carnival grab-bag: Starburst galaxies, BL Lacertae galaxies, Seyfert galaxies (types I and II), blazars, N-galaxies, LINERS, infrared galaxies, radio galaxies, and of course, the royalty of active galaxies—quasars. The extraordinary luminosities of these elite galaxies derive from mysterious activity within a small region buried deep within their nucleus.

Quasars, discovered in the early 1960s, are the most exotic of them all. Some are a thousand times as luminous as our own Milky Way galaxy, yet their energy hails from a region that would fit comfortably within the planetary orbits of our solar system. Curiously, none are nearby. The closest one sits about 1.5 billion light-years away—its light has been traveling for 1.5 billion years to reach us. And most quasars hail from beyond 10 billion light-years. Possessed of small size and extreme distance, on photographs one can hardly distinguish them from the pointlike images left by local stars in our own Milky Way, leaving visible-light telescopes quite useless as tools of discovery. The earliest quasars were, in fact, discovered using radio telescopes. Since stars do not emit copious amounts of radio waves, these radio-bright objects were a new class of something or other, masquerading as a star. In the we-call-them-as-we-see-them tradition among astrophysicists, these objects were dubbed Quasi-Stellar Radio Sources, or more affectionately, "quasars."

What manner of beast are they?

One's ability to describe and understand a new phenomenon is always limited by the contents of the prevailing scientific and technological toolbox. An eighteenth-century person who was briefly, but unwittingly, thrust into the twentieth century would return and describe a car as a horse-drawn carriage without the horse and a lightbulb as a candle without the flame. With no knowledge of internal combustion engines or electricity, a true understanding would be remote indeed. With that as a disclaimer, allow me to declare that we think we understand the basics principles of what drives a quasar. In what has come to be known as the "standard model," black holes have been implicated as the engine of quasars and of all active galaxies. Within a black hole's boundary of space and time—its event horizon—the concentration of matter is so great that the velocity needed to escape exceeds the speed of light. Since the speed of light is a universal limit, when you fall into a black hole, you fall in for good, even if you're made of light.

HOW, MIGHT YOU ASK, can something that emits no light power something that emits more light than anything else in the universe?

In the late 1960s and 1970s, it didn't take long for astrophysicists to discover that the exotic properties of black holes made remarkable additions the theorists' toolbox. According to some well-known laws of gravitational physics, as gaseous matter funnels toward a black hole, the matter must heat up and radiate profusely before it descends through the event horizon. The energy comes from the efficient conversion of gravity's potential energy into heat.

While not a household notion, we have all seen gravitational potential energy get converted at some time in our terrestrial lives. If you have ever dropped a dish to the floor and broken it, or if you have ever nudged something out the window that splattered on the ground below, then you understand the power of gravitational potential energy. It's simply untapped energy endowed by an object's distance from wherever it might hit if it fell. When objects fall, they normally gain speed. But if something stops the fall, all the energy the object had gained converts to the kind of energy that breaks or splatters things. Therein is the real reason why you are more likely to die if you jump off a tall building instead of a short building.

If something prevents the object from gaining speed yet the object continues to fall, then the converted potential energy reveals itself some other way—usually in the form of heat. Good examples include space vehicles and meteors when they heat up while descending through Earth's atmosphere: they want to speed up, but air resistance prevents it. In a now-famous experiment, the nineteenth-century English physicist James Joule created a device that stirred a jar of water with rotating paddles by the action of falling weights. The potential energy of the weights was transferred into the water and successfully raised its temperature. Joule describes his effort:

> The paddle moved with great resistance in the can of water, so that the weights (each of four pounds) descended at the slow rate of about one foot per second. The height of the pulleys from the ground was twelve yards, and consequently, when the weights had descended

through that distance, they had to be wound up again in order to renew the motion of the paddle. After this operation had been repeated sixteen times, the increase of the temperature of the water was ascertained by means of a very sensible and accurate thermometer. . . . I may therefore conclude that the existence of an equivalent relation between heat and the ordinary forms of mechanical power is proved. . . . If my views are correct, the temperature of the river Niagara will be raised about one fifth of a degree by its fall of 160 feet. (Shamos 1959, p. 170)

Joule's thought-experiment refers, of course, to the great Niagara Falls. But had he known of black holes, he might have said instead, "If my views are correct, the temperature of gas funneled toward a black hole will be raised a million degrees by its fall of a billion miles."

AS YOU MIGHT suspect, black holes enjoy a prodigious appetite for stars that wander too close. A paradox of galactic engines is that their black holes must eat to radiate. The secret to powering the galactic engine lies in a black hole's ability to ruthlessly and gleefully rip apart stars before they cross the event horizon. The tidal forces of gravity for a black hole elongate the otherwise spherical stars in much the same way that the Moon's tidal forces elongate Earth's oceans to create high and low oceanic tides. Gas that was formerly part of stars (and possibly ordinary gas clouds) cannot simply gain speed and fall in; the gas of previously shredded stars impedes wanton free fall down the hole. The result? A star's gravitational potential energy gets converted to prodigious levels of heat and radiation. And the higher the gravity of your target, the more gravitational potential energy was available to convert.

Faced with the proliferation of words to describe oddball galaxies, the late Gerard de Vaucouleurs (1983), a consummate morphologist, was quick to remind the astronomical community that a car that has been wrecked does not all of a sudden become a different kind of car. This car-wreck philosophy has led to a standard model of active galaxies that largely unifies the zoo. The model is endowed with enough tweakable parts to explain most of the basic, observed features. For example, the funneling gas often forms an opaque rotating disk before it descends through the event horizon. If the outward flow of radiation cannot penetrate the disk of accreted gas, then radiation will fly out from above and below the disk to create titanic jets of matter and energy. The observed properties of the galaxy will be different if the galaxy's jet happens to be pointing toward you or sideways to you—or if the ejected material moves slowly or at speeds close to the speed of light. The thickness and chemical composition of the disk will also influence its appearance as well as the rate at which stars are consumed.

To feed a healthy quasar requires that its black hole eat up to ten stars per year. Other less-active galaxies from our carnival shred many fewer stars per year. For many quasars, their luminosity varies on time scales of days and even hours. Allow me to impress you with how extraordinary this is. If the active part of a quasar were the size of the Milky Way (100,000 light-years across) and if it all decided to brighten at once, then you would first learn about it from the side of the galaxy that was closest to you, and then 100,000 years later the last part of the galaxy's light would reach you. In other words, it would take 100,000 years for you to observe the quasar brightening fully. For a quasar to brighten within hours means that the dimensions of the engine cannot be greater than light-hours across. How big is that? About the size of the solar system.

With a careful analysis of the light fluctuations in all bands, a crude, but informative three-dimensional structure can be deduced for the surrounding material. For example, the luminosity in x-rays might vary over a time scale of hours but the red light might vary over weeks. The comparison allows you to conclude that the red light-emitting part of the active galaxy is much larger than the x-ray emitting part. This exercise can be invoked for many bands of light to derive a remarkably complete picture of the system.

If most of this action takes place during the early universe in distant quasars, then why does it no longer happen? Why are there no local quasars? Do dead quasars lurk under our noses?

Good explanations are available. The most obvious is that the core of local galaxies ran out of stars to feed the engine, having vacuumed up all stars whose orbits came too close to the black hole. No more food, no more prodigious regurgitations.

A more interesting shut-off mechanism comes from what happens to the tidal forces as the black hole's mass (and event horizon) grows and grows. As we will see later in this section, tidal forces have nothing to do with the total gravity felt by an object—what matters is the difference in gravity across it, which increases dramatically as you near an object's center. So large, high-mass black holes actually exert lower tidal forces than the smaller, low-mass black holes. No mystery here. The Sun's gravity on Earth dwarfs that of the Moon's on Earth yet the proximity of the Moon enables it to exert considerably higher tidal forces at our location, a mere 240,000 miles away.

It's possible, then, for a black hole to eat so much that its event horizon grows so large that its tidal forces are no longer sufficient to shred a star. When this happens, all of the star's gravitational potential energy converts to the star's speed and the star gets eaten whole as it plunges past the event horizon. No more conversion to heat and radiation. This shut-off valve kicks in for a black hole of about a billion times the mass of the Sun.

These are powerful ideas that do indeed offer a rich assortment of explanatory tools. The unified picture predicts that quasars and other active galaxies are just early chapters in the life of a galaxy's nucleus. For this to be true, specially exposed images of quasars should reveal the surrounding fuzz of a host galaxy. The observational challenge is similar to that faced by solar system hunters who try to detect planets hidden in the glare of their host star. The quasar is so much brighter than the surrounding galaxy that special masking techniques must be used to detect anything other than the quasar itself. Sure enough, nearly all high-resolution images of quasars reveal surrounding galaxy fuzz. The several exceptions of uncloaked quasars continue to confound the expectations of the standard model. Or is it that the host galaxies simply fall below the detection limits?

The unified picture also predicts that quasars would eventually shut themselves off. Actually, the unified picture must predict this because the absence of nearby quasars requires it. But it also means that black holes in galactic nuclei should be common, whether or not the galaxy has an active nucleus. Indeed, the list of nearby galaxies that contain dormant supermassive black holes in their nuclei is growing longer by the month and includes the Milky Way. Their existence is betrayed through the astronomical speeds that stars achieve as they orbit close (but not too close) to the black hole itself.

Fertile scientific models are always seductive, but one should occasionally ask whether the model is fertile because it captures some deep truths about the universe or because it was constructed with so many tunable variables that you can explain anything at all. Have we been sufficiently clever today, or are we missing a tool that will be invented or discovered tomorrow? The English physicist Dennis Sciama knew this dilemma well when he noted:

Since we find it difficult to make a suitable model of a certain type, Nature must find it difficult too. This argument neglects the possibility that Nature may be cleverer than we are. It even neglects the possibility that we may be cleverer to-morrow than we are to-day. (1971, p. 80)

THIRTY-TWO

KNOCK 'EM DEAD

Ever since people discovered the bones of extinct dinosaurs, scientists have proffered no end of explanations for the disappearance of the hapless beasts. Maybe a torrid climate dried up the available sources of water, some say. Maybe volcanoes covered the land in lava and poisoned the air. Maybe Earth's orbit and axis tilt brought on a relentless ice age. Maybe too many early mammals dined on too many dino eggs. Or maybe the meat-eating dinosaurs ate up all the vegetarian ones. Maybe the need to find water led to massive migrations that rapidly spread diseases. Maybe the real problem was a reconfiguration of landmasses, caused by tectonic shifts.

All these crises have one thing in common: the scientists who came up with them were well trained in the art of looking down.

Other scientists, however, trained in the art of looking up, began to make connections between Earth's surface features and the visits of vagabonds from outer space. Maybe meteor impacts generated some of those features, such as Barringer Crater, that famous, mile-wide, bowl-shaped depression in the Arizona desert. In the 1950s, the American geologist Eugene M. Shoemaker and his associates discovered a kind of rock that forms only under short-lived, but extremely high, pressure—just what a fast-moving meteor would do. Geologists could finally agree that an impact caused the bowl (now sensibly called Meteor Crater), and Shoemaker's discovery resurrected the nineteenth-century concept of catastrophism—the idea that changes to our planet's skin can be caused by brief, powerful, destructive events.

Once the gates of speculation opened, people began to wonder whether the dinosaurs might have disappeared at the hands of a similar, but bigger, assault. Meet iridium: a metal rare on Earth but common in metallic meteorites and conspicuous in a 65-million-year-old layer of clay that appears at scores of sites around the world. That clay, dating to about the same time as the dinos checked out, marks the crime scene: the end of the Cretaceous. Now meet Chicxulub Crater, a 200-kilometer-wide depression at the edge of Mexico's Yucatan Peninsula. It, too, is about 65 million years old. Computer simulations of climate change make it clear that any impact that could make that crater would thrust enough of Earth's crust in the stratosphere that global climatic catastrophe would ensue. Who could ask for anything more? We've got the perpetrator, the smoking gun, and a confession.

Case closed.

Or is it?

Scientific inquiry shouldn't stop just because a reasonable explanation has apparently been found. Some paleontologists and geologists remain skeptical about assigning Chicxulub the lion's share—or even a substantial share—of responsibility for the dinos' departure. Some think Chicxulub may have significantly predated the extinction. Furthermore, Earth was volcanically busy at about that time. Plus, other waves of extinction have swept across Earth without leaving craters and rare cosmic metals as calling cards. And not all bad things that arrive from space leave a crater. Some explode in midair and never make it to Earth's surface.

So, besides impacts, what else might a restless cosmos have in store for us? What else could the universe send our way that might swiftly unravel the patterns of life on Earth?

SEVERAL SWEEPING EPISODES of mass extinction have punctuated the past half-billion years on Earth. The biggest are the Ordovician, about 440 million years ago; the Devonian, about 370 million; the Permian, about 250 million; the Triassic, about 210 million; and, of course, the Cretaceous, about 65 million. Lesser extinction episodes have taken place as well, at timescales of tens of millions of years.

Some investigators pointed out that, on average, an episode of note takes place every 25 million years or so. People who spend most of their time looking up are comfortable with phenomena that repeat at long intervals, and so astrophysicists decided it was our turn to name some killers.

Let's give the Sun a dim and distant companion star, a few up-lookers said in the 1980s. Let's declare its orbital period to be about 25 million years and its orbit to be extremely elongated, so that it spends most of its time too far from Earth to be detected. This companion would discombobulate the Sun's distant reservoir of comets whenever it passed through their neighborhood. Legions of comets would jiggle loose from their stately orbits in the outer solar system, and the rate of impacts on Earth's surface would vastly increase.

Therein was the genesis of Nemesis, the name given to this hypothetical killer star. Subsequent analyses of the extinction episodes have convinced most experts that the average time between catastrophes varies too greatly to signify anything truly periodic. But for a few years the idea was big news.

Periodicity wasn't the only intriguing idea about death from outer space. Pandemics were another. The late English astrophysicist Sir Fred Hoyle and his longtime collaborator Chandra Wickramasinghe, now at Cardiff University in Wales, pondered whether Earth might occasionally pass through an interstellar cloud laden with microorganisms or be on the receiving end of similarly endowed dust from a passing comet. Such an encounter might give rise to a fast-spreading illness, they suggested. Worse yet, some of the giant clouds or dust trails might be real killers—

bearing viruses with the power to infect and destroy a wide range of species. Of the many challenges to making this idea work, nobody knows how an interstellar cloud could manufacture and carry something as complex as a virus.

You want more? Astrophysicists have imagined a nearly endless spectrum of awesome catastrophes. Right now, for instance, the Milky Way galaxy and the Andromeda galaxy, a near twin of ours 2.4 million light-years up the road, are falling toward each other. As discussed earlier, in about 7 billion years they may collide, causing the cosmic equivalent of a train wreck. Gas clouds would slam into one another; stars would be cast hither and yon. If another star swung close enough to confound our gravitational allegiance to the Sun, our planet could get flung out of the solar system, leaving us homeless in the dark.

That would be bad.

Two billion years before that happens, however, the Sun itself will swell up and die of natural causes, engulfing the inner planets— including Earth—and vaporizing all their material contents.

That would be worse.

And if an interloping black hole comes too close to us, it will dine on the entire planet, first crumbling the solid Earth into a rubble pile by virtue of its unstoppable tidal forces. The remains would then be extruded though the fabric of space-time, descending as a long string of atoms through the black hole's event horizon, down to its singularity.

But Earth's geologic record never mentions any early close encounters with a black hole—no crumbling, no eating. And given that we expect a vanishingly low number of neighborhood black holes, I'd say we have more pressing issues of survival before us.

HOW ABOUT GETTING fried by waves of high-energy electromagnetic radiation and particles, spewed across space by an exploding star?

Most stars die a peaceful death, gently shedding their outer gases into interstellar space. But one in a thousand—the star whose mass is greater than about seven or eight times that of the Sun—dies in a violent, dazzling explosion called a supernova. If we found ourselves within 30 light-years of one of those, a lethal dose of cosmic rays—high-energy particles that shoot across space at almost the speed of light—would come our way.

The first casualties would be ozone molecules. Stratospheric ozone (O_3) normally absorbs damaging ultraviolet radiation from the Sun. In so doing, the radiation breaks the ozone molecule apart into oxygen (O) and molecular oxygen (O_2). The newly freed oxygen atoms can then join forces with other oxygen molecules, yielding ozone once again. On a normal day, solar ultraviolet rays destroy Earth's ozone at the same rate as the ozone gets replenished. But an overwhelming high-energy assault on our stratosphere would destroy the ozone too fast, leaving us all in desperate need of sunblock.

Once the first wave of cosmic rays took out our defensive ozone, the Sun's ultraviolet would sail clear down to Earth's surface, splitting oxygen and nitrogen molecules as it went. For the birds, mammals, and other residents of Earth's surface and airspace, that would be unpleasant news indeed. Free oxygen atoms and free nitrogen atoms would readily combine. One product would be nitrogen dioxide, a component of smog, which would darken the atmosphere and cause the temperature to plummet. A new ice age might dawn even as the ultraviolet rays slowly sterilized Earth's surface.

BUT THE ULTRAVIOLET blasted in every direction by a supernova is just a mosquito bite compared to the gamma rays let loose from a hypernova.

At least once a day, a brief burst of gamma rays—the highest of high-energy radiation—unleashes the energy of a thousand supernovas somewhere in the cosmos. Gamma-ray bursts were accidentally discovered in the 1960s by U.S. Air Force satellites, launched to detect radiation from any clandestine nuclear-weapons tests the Soviet Union might have conducted in violation of the 1963 Limited Test Ban Treaty. What the satellites found instead were signals from the universe itself.

At first nobody knew what the bursts were or how far away they took place. Instead of clustering along the plane of the Milky Way's main disk of stars and gas, they came from every direction on the sky—in other words, from the entire cosmos. Yet surely they had to be happening nearby, at least within a galactic diameter or so from us. Otherwise, how was it possible to account for all the energy they registered here on Earth?

In 1997 an observation made by an orbiting Italian x-ray telescope settled the argument: gamma-ray bursts are extremely distant extragalactic events, perhaps signaling the explosion of a single supermassive star and the attendant birth of a black hole. The telescope had picked up the x-ray "afterglow" of a now-famous burst, GRB 970228. But the x-rays were "redshifted." Turns out, this handy feature of light and the expanding universe enables astrophysicists to make a fairly accurate determination of distance. The afterglow of GRB 970228, which reached Earth on February 28, 1997, was clearly coming from halfway across the universe, billions of light-years away. The following year Bohdan Paczynski, a Princeton astrophysicist, coined the term "hypernova" to describe the source of such bursts. Personally, I would have voted for "super-duper supernova."

A hypernova is the one supernova in 100,000 that produces a gamma-ray burst, generating in a matter of moments the same amount of energy as our Sun would emit if it shone at its present output for a trillion years. Barring the influence of some undiscovered law of physics, the only way to achieve the measured energy is to beam the total output of the explosion in a narrow ray—much the way all the light from a flashlight bulb gets channeled by the flashlight's parabolic mirror into one strong, forward-pointing beam. Pump a supernova's power through a narrow beam, and anything in the beam's path will get the full brunt of the explosive energy. Meanwhile, whoever does not fall in the beam's path remains oblivious. The narrower the beam, the more intense the flux of its energy and the fewer the cosmic occupants who will see it.

What gives rise to these laserlike beams of gamma rays? Consider the original supermassive star. Not long before its death from fuel starvation, the star jettisons its outer layers. It becomes cloaked in a vast, cloudy shell, possibly augmented by pockets of gas left over from the cloud that originally spawned the star. When the star finally collapses and explodes, it releases stupendous quantities of matter and prodigious quantities of energy. The first assault of matter and energy punches through weak points in the shell of gas, enabling the succeeding matter and energy to funnel through that same point. Computer models of this complicated scenario suggest that the weak points are typically just above the north and south poles of the original star. When seen from beyond the shell, two powerful beams travel in opposite directions, headed toward all gamma-ray detectors (test-ban-treaty detectors or otherwise) that happen to lie in their path.

Adrian Melott, an astronomer at the University of Kansas, and an interdisciplinary crew of colleagues assert that the Ordovician extinction may well have been caused by a face-to-face encounter with a nearby gamma-ray burst. A quarter of Earth's families of organisms perished at that time. And nobody has turned up evidence of a meteor impact contemporary with the event.

WHEN YOU'RE A hammer (as the saying goes), all your problems look like nails. If you're a meteorite expert pondering the sudden extinction of boatloads of species, you'll want to say an impact did it. If you're an igneous petrologist, volcanoes did it. If you're into spaceborne bioclouds, an interstellar virus did it. If you're a hypernova expert, gamma rays did it.

No matter who is right, one thing is certain: whole branches in the tree of life can go extinct almost instantly.

Who survives these assaults? It helps to be small and meek. Microorganisms tend to do well in the face of adversity. More important, it helps if you live where the Sun don't shine—on the bottom of the ocean, in the crevices of buried rocks, in the clays and soils of farms and forests. The vast underground biomass survives. It is they who inherit the Earth again, and again, and again.

THIRTY-THREE

DEATH BY BLACK HOLE

Without a doubt, the most spectacular way to die in space is to fall into a black hole. Where else in the universe can you lose your life by being ripped apart atom by atom?

Black holes are regions of space where the gravity is so high that the fabric of space and time has curved back on itself, taking the exit doors with it. Another way to look at the dilemma: the speed required to escape a black hole is greater than the speed of light itself. As we saw back in Section 3, light travels at exactly 299,792,458 meters per second in a vacuum and is the fastest stuff in the universe. If light cannot escape, then neither can you, which is why, of course, we call these things black holes.

All objects have escape speeds. Earth's escape speed is a mere 11 kilometers per second, so light escapes freely, as would anything else launched faster than 11 kilometers per second. Please tell all those people who like to proclaim, "What goes up must come down!" that they are misinformed.

Albert Einstein's general theory of relativity, published in 1916, provides the insight to understand the bizarre structure of space and time in a high-gravity environment. Later research by the American physicist John A. Wheeler, and others, helped to formulate a vocabulary as well as the mathematical tools to describe and predict what a black hole will do to its surroundings. For example, the exact boundary between where light can and cannot escape, which also separates what's in the universe and what's forever lost to the black hole, is poetically known as the "event horizon." And by convention, the size of a black is the size of its event horizon, which is a clean quantity to calculate and to measure. Meanwhile, the stuff within the event horizon has collapsed to an infinitesimal point at the black hole's center. So black holes are not so much deadly objects as they are deadly regions of space.

Let's explore in detail what black holes do to a human body that wanders a little too close.

If you stumbled upon a black hole and found yourself falling feet-first toward its center, then as you got closer, the black hole's force of gravity would grow astronomically. Curiously, you would not feel this force at all because, like anything in free fall, you are weightless. What you do feel, however, is something far more sinister. While you fall, the black hole's force of gravity at your two feet, they being closer to the black hole's center, accelerates them faster than does the weaker force of gravity at your head. The difference between the two is known officially as the tidal force, which grows precipitously as you draw nearer to the black hole's center. For Earth, and for most cosmic places, the tidal force across the length of your body is minuscule and goes unnoticed. But in your feet-first fall toward a black hole the tidal forces are all you notice.

If you were made of rubber then you would just stretch in response. But humans are composed of other materials such as bones and muscles and organs. Your body would stay whole until the instant the tidal force exceeded your body's molecular bonds. (If the Inquisition had access to black holes, this, instead of the rack, would surely have become the stretching device of choice.)

That's the gory moment when your body snaps into two segments, breaking apart at your midsection. Upon falling further, the difference in gravity continues to grow, and each of your two body segments snaps into two segments. Shortly thereafter, those segments each snap into two segments of their own, and so forth, and so forth, bifurcating your body into an ever-increasing number of parts: 1, 2, 4, 8, 16, 32, 64, 128, etc. After you've been ripped into shreds of organic molecules, the molecules themselves begin to feel the continually growing tidal forces. Eventually, they too snap apart, creating a stream of their constituent atoms. And then, of course, the atoms themselves snap apart, leaving an unrecognizable parade of particles that, minutes earlier, had been you.

But there is more bad news.

All parts of your body are moving toward the same spot—the black hole's center. So while you're getting ripped apart head to toe, you will also extrude through the fabric of space and time, like toothpaste squeezed through a tube.

To all the words in the English language that describe ways to die (e.g., homicide, suicide, electrocution, suffocation, starvation) we add the term "spaghettification."

AS A BLACK HOLE eats, its diameter grows in direct proportion to its mass. If, for example, a black hole eats enough to triple its mass, then it will have grown three times as wide. For this reason, black holes in the universe can be almost any size, but not all of them will spaghettify you before you cross the event horizon. Only "small" black holes will do that. Why? For a graphic, spectacular death, all that matters is the tidal force. And as a general rule, the tidal force on you is greatest if your size is large compared with your distance to the center of the object.

In a simple but extreme example, if a six-foot man (who is not otherwise prone to ripping apart) falls feet-first toward a six-foot black hole, then at the event horizon, his head is twice as far away from the black hole's center as his feet. Here, the difference in the force of gravity from his feet to his head would be very large. But if the black hole were 6,000 feet across, then the same man's feet would be only one-tenth of 1 percent closer to the center than his head, and the difference in gravity—the tidal force—would be correspondingly small.

Equivalently, one can ask the simple question: How quickly does the force of gravity change as you draw nearer to an object? The equations of gravity show that gravity changes more and more swiftly as you near the center of an object. Smaller black holes allow you to get much closer to their centers before you enter their event horizons, so the change of gravity over small distances can be devastating to fallers-in.

A common variety of black hole contains several times the mass of the Sun, but packs it all within an event horizon only about a dozen miles across. These are what most astronomers discuss in casual conversations on the subject. In a fall toward this beast, your body would begin to break apart within 100 miles of the center. Another common variety of black hole reaches a billion times the mass of the Sun and is contained within an event horizon that is nearly the size of the entire solar system. Black holes such as these are what lurk in the centers of galaxies. While their total gravity is monstrous, the difference in gravity from your head to your toes near their event horizons is relatively small. Indeed, the tidal force can be so weak that you will likely fall through the event horizon in one piece—you just wouldn't ever be able to come back out and tell anybody about your trip. And when you do finally get ripped apart, deep within the event horizon, nobody outside the hole will be able to watch.

As far as I know, nobody has ever been eaten by a black hole, but there is compelling evidence to suggest that black holes in the universe routinely dine upon wayward stars and unsuspecting gas clouds. As a cloud approaches a black hole, it hardly ever falls straight in. Unlike your choreographed feet-first fall, a gas cloud is typically drawn into orbit before it spirals to its destruction. The parts of the cloud that are closer to the black hole will orbit faster than the parts that are farther away. Known as differential rotation, this simple shearing can have extraordinary astrophysical consequences. As the cloud layers spiral closer to the event horizon they heat up, from internal friction, to upwards of a million degrees— much hotter than any known star. The gas glows blue-hot as it becomes a copious source of ultraviolet and x-ray energy. What started as an isolated, invisible black hole (minding its own business) has now become an invisible black hole encircled by a gaseous speedway, ablaze with high-energy radiation.

Since stars are 100 percent certified balls of gas, they are not immune from the fate that greeted our hapless clouds. If one star in a binary system becomes a black hole, then the black hole does not get to eat until late in the companion star's life, when it swells to become a red giant. If the red giant grows large enough, then it will ultimately get flayed, as the black hole peels and eats the star, layer by layer. But for a star that just happens to wander into the neighborhood, tidal forces will initially stretch it, but eventually, differential rotation will shear the star into a friction-heated disk of highly luminous gas.

Whenever a theoretical astrophysicist needs an energy source in a tiny space to explain a phenomenon, well-fed black holes become prime ammunition. For example, as we saw earlier, the distant and mysterious quasars wield hundreds or thousands of times the luminosity of the entire Milky Way galaxy. But their energy emanates primarily from a volume that is not much larger than our solar system. Without invoking a supermassive black hole as the quasar's central engine, we are at a loss to find an alternative explanation.

We now know that supermassive black holes are common in the cores of galaxies. For some galaxies, a suspiciously high luminosity in a suspiciously small volume provides the needed smoking gun, but the actual luminosity depends heavily on whether stars and gas are available for the black hole to shear them apart. Other galaxies may have one too, in spite of an unremarkable central luminosity. These black holes may have already eaten all the surrounding stars and gas, leaving no evidence behind. But stars near the center, in close orbit to the black hole (not too close to be consumed), will have sharply increased speeds.

These speeds, when combined with the stars' distance from the center of the galaxy, are a direct measure of the total mass contained within their orbits. Armed with these data, we can use the back of an envelope to calculate whether the attracting central mass is, indeed, concentrated enough to be a black hole. The largest known black holes are typically a billion solar masses, such as what lurks within the titanic elliptical galaxy M87, the largest in the Virgo Cluster of galaxies. Far down the list, but still large, is the 30-million solar mass black hole in the center of the Andromeda galaxy, our near neighbor in space.

Beginning to feel "black hole envy"? You are entirely justified: the one in the Milky Way's center checks in at a mere 4-million solar masses. But no matter the mass, death and destruction are their business.

SCIENCE
AND
CULTURE

THE RUFFLED INTERFACE BETWEEN
COSMIC DISCOVERY AND THE
PUBLIC'S REACTION TO IT

THINGS PEOPLE SAY

ristotle once declared that while the planets moved against the background stars, and while shooting stars, comets, and eclipses represented intermittent variability in the atmosphere and the heavens, the stars themselves were fixed and unchanging on the sky and that Earth was the center of all motion in the universe. From our enlightened perch, 25 centuries later, we chuckle at the folly of these ideas, but the claims were the consequence of legitimate, albeit simple, observations of the natural world.

Aristotle also made other kinds of claims. He said that heavy things fall faster than light things. Who could argue against that? Rocks obviously fall to the ground faster than tree leaves. But Aristotle went further and declared that heavy things fall faster than light things in direct proportion to their own weight, so that a 10-pound object would fall ten times faster than a 1-pound object.

Aristotle was badly mistaken.

To test him, simply release a small rock and a big rock simultaneously from the same height. Unlike fluttering leaves, neither rock will be much influenced by air resistance and both will hit the ground at the same time. This experiment does not require a grant from the National Science Foundation to execute. Aristotle could have performed it but didn't. Aristotle's teachings were later adopted into the doctrines of the Catholic Church. And through the Church's power and influence Aristotelian philosophies became lodged in the common knowledge of the Western world, blindly believed and repeated. Not only did people repeat to others that which was not true, but they also ignored things that clearly happened but were not supposed to be true.

When scientifically investigating the natural world, the only thing worse than a blind believer is a seeing denier. In A.D. 1054, a star in the constellation Taurus abruptly increased in brightness by a factor of a million. The Chinese astronomers wrote about it. Middle Eastern astronomers wrote about it. Native Americans of what is now the southwestern United States made rock engravings of it. The star became bright enough to be plainly visible in the daytime for weeks, yet we have no record of anybody in all of Europe recording the event. (The bright new star in the sky was actually a supernova explosion that occurred in space some 7,000 years earlier but its light had only just reached Earth.) True, Europe was in the Dark Ages, so we cannot expect that acute data-taking skills were common, but cosmic events that were "allowed" to happen were routinely recorded. For example, 12 years later, in 1066, what ultimately became known as Halley's comet was seen and duly depicted—complete with agape onlookers—in a section of the famous Bayeux tapestry, circa 1100. An exception indeed. The Bible says the stars don't change. Aristotle said the stars don't change. The Church, with its unmatched authority, declares the star's don't change. The population then falls victim to a collective delusion that was stronger than its members' own powers of observation.

We all carry some blindly believed knowledge because we cannot realistically test every statement uttered by others. When I tell you that the proton has an antimatter counterpart (the antiproton), you would need $1 billion worth of laboratory apparatus to verify my statement. So it's easier to just believe me and trust that, at least most of the time, and at least with regard to the astrophysical world, I know what I am talking about. I don't mind if you remain skeptical. In fact, I encourage it. Feel free to visit your nearest particle accelerator to see antimatter for yourself. But how about all those statements that don't require fancy apparatus to prove? One would think that in our modern and enlightened culture, popular knowledge would be immune from falsehoods that were easily testable.

It is not.

Consider the following declarations. The North Star is the brightest star in the nighttime sky. The Sun is a yellow star. What goes up must come down. On a dark night you can see millions of stars with the unaided eye. In space there is no gravity. A compass points north. Days get shorter in the winter and longer in the summer. Total solar eclipses are rare.

Every statement in the above paragraph is false.

Many people (perhaps most people) believe one or more of these statements and spread them to others even when a firsthand demonstration of falsehood is trivial to deduce or obtain. Welcome to my things-people-say rant:

The North Star is not the brightest star in the nighttime sky. It's not even bright enough to earn a spot in the celestial top 40. Perhaps people equate popularity with brightness. But when gazing upon the northern sky, three of the seven stars of the Big Dipper, including its "pointer" star, are brighter than the North Star, which is parked just three fist-widths away. There is no excuse.

And I don't care what else anyone has ever told you, the Sun is white, not yellow. Human color perception is a complicated business, but if the Sun were yellow, like a yellow lightbulb, then white stuff such as snow would reflect this light and appear yellow—a snow condition confirmed to happen only near fire hydrants. What could lead people to say that the Sun is yellow? In the middle of the day, a glance at the Sun can damage your eyes. Near sunset, however, with the Sun low on the horizon and when the atmospheric scattering of blue light is at its greatest, the Sun's intensity is significantly diminished. The blue light from the Sun's spectrum, lost to the twilight sky, leaves behind a yellow-orange-red hue for the Sun's disk. When people glance at this color-corrupted setting Sun, their misconceptions are fueled.

What goes up need not come down. All manner of golf balls,

flags, automobiles, and crashed space probes litter the lunar surface. Unless somebody goes up there to bring them back, they will never return to Earth. Not ever. If you want to go up and not come down, all you need to do is travel at any speed faster than about seven miles per second. Earth's gravity will gradually slow you down but it will never succeed in reversing your motion and forcing you back to Earth.

Unless your eyes have pupils the size of binocular lenses, no matter your seeing conditions and no matter your location on Earth, you will not resolve any more than about five or six thousand stars in the entire sky out of the 100 billion (or so) stars of our Milky Way galaxy. Try it one night. Things get much, much worse when the Moon is out. And if the Moon happens to be full, it will wash out the light of all but the brightest few hundred stars.

During the Apollo space program, while one of the missions was en route to the Moon, a noted television news anchor announced the exact moment when the "astronauts left the gravitational field of Earth." Since the astronauts were still on their way to the Moon, and since the Moon orbits Earth, then Earth's gravity must extend into space *at least as far as the Moon*. Indeed, Earth's gravity, and the gravity of every other object in the universe, extends without limit—albeit with ever-diminishing strength. Every spot in space is teeming with countless gravitational tugs in the direction of every other object in the universe. What the announcer meant was that the astronauts crossed the point in space where the force of the Moon's gravity exceeds the force of Earth's gravity. The whole job of the mighty three-stage *Saturn V* rocket was to endow the command module with enough initial speed to just reach this point in space because thereafter you can passively accelerate toward the Moon—and they did. Gravity is everywhere.

Everybody knows that when it comes to magnets, opposite poles attract while similar poles repel. But a compass needle is designed so that the half that has been magnetized "North" points to Earth's magnetic north pole. The only way a magnetized object can align its north half to Earth's magnetic north pole is if Earth's magnetic north pole is actually in the south and the magnetic south pole is actually in the north. Furthermore, there is no particular law of the universe that requires the precise alignment of an object's magnetic poles with its geographic poles. On Earth the two are separated by about 800 miles, which makes navigation by compass a futile exercise in northern Canada.

Since the first day of winter is the shortest "day" of the year, then every succeeding day in the winter season must get longer and longer. Similarly, since the first day of summer is the longest "day" of the year, then every succeeding day in the summer must get shorter and shorter. This is, of course, the opposite of what is told and retold.

On average, every couple years, somewhere on Earth's surface, the Moon passes completely in front of the Sun to create a total solar eclipse. This event is more common than the Olympics, yet you don't read newspaper headlines declaring "a rare Olympics will take place this year." The perceived rarity of eclipses may derive from a simple fact: for any chosen spot on Earth, you can wait up to a half-millennium before you see a total solar eclipse. True, but lame as an argument because there are spots on Earth (like the middle of the Sahara Desert or any region of Antarctica) that have never, and will not likely ever, host the Olympics.

Want a few more? At high noon, the Sun is directly overhead. The Sun rises in the east and sets in the west. The Moon comes out at night. On the equinox there are 12 hours of day and 12 hours of night. The Southern Cross is a beautiful constellation. All of these statements are wrong too.

There is no time of day, nor day of the year, nor place in the continental United States where the Sun ascends to directly overhead. At "high noon," straight vertical objects cast no shadow. The only people on the planet who see this live between 23.5 degrees south latitude and 23.5 degrees north latitude. And even in that zone, the Sun reaches directly overhead on only two days per year. The concept of high noon, like the brightness of the North Star and the color of the Sun, is a collective delusion.

For every person on Earth, the Sun rises due east and sets due west on only two days of the year: the first day of spring and the first day of fall. For every other day of the year, and for every person on Earth, the Sun rises and sets someplace else on the horizon. On the equator, sunrise varies by 47 degrees across the eastern horizon. From the latitude of New York City (41 degrees north—the same as that of Madrid and Beijing) the sunrise spans more than 60 degrees. From the latitude of London (51 degrees north) the sunrise spans nearly 80 degrees. And when viewed from either the Arctic or Antarctic circles, the Sun can rise due north and due south, spanning a full 180 degrees.

The Moon also "comes out" with the Sun in the sky. By invoking a small extra investment in your skyward viewing (like looking up in broad daylight) you will notice that the Moon is visible in the daytime nearly as often as it is visible at night.

The equinox does not contain exactly 12 hours of day and 12 hours of night. Look at the sunrise and sunset times in the newspaper on the first day of either spring or fall. They do not split the day into two equal 12-hour blocks. In all cases, daytime wins. Depending on your latitude, it can win by as few as seven minutes at the equator up to nearly half an hour at the Arctic and Antarctic circles. Who or what do we blame? Refraction of sunlight as it passes from the vacuum of interplanetary space to Earth's atmosphere enables an image of the Sun to appear above the horizon several minutes before the actual Sun has actually risen. Equivalently, the actual Sun has set several minutes before the Sun that you see. The convention is to measure sunrise by using the upper edge of the Sun's disk as it peeks above the horizon; similarly, sunset is measured by using the upper edge of the Sun's disk as it sinks below the horizon. The problem is that these two "upper edges" are on opposite halves of the Sun thereby providing an extra solar width of light in the sunrise/sunset calculation.

The Southern Cross gets the award for the greatest hype among all eighty-eight constellations. By listening to Southern Hemisphere people talk about this constellation, and by listening to songs written about it, and by noticing it on the national flags of Australia, New Zealand, Western Samoa, and Papua New Guinea, you would think we in the North were somehow deprived. Nope. Firstly, one needn't travel to the Southern Hemisphere to see the Southern Cross. It's plainly visible (although low in the sky) from as far north as Miami, Florida. This diminutive constellation is the smallest in the sky— your fist at arm's length would eclipse it completely. Its shape isn't very interesting either. If you were to draw a rectangle using a connect-the-dots method you would use four stars. And if you were to draw a cross you would presumably include a fifth star in the middle to indicate the cross-point of the two beams. But the Southern Cross is composed of only four stars, which more accurately resemble a kite or a crooked box. The constellation lore of Western cultures owes its origin and richness to centuries of Babylonian, Chaldean, Greek, and Roman imaginations. Remember, these are the same imaginations that gave rise to the endless dysfunctional social lives of the gods and goddesses. Of course, these were all Northern Hemisphere civilizations, which means the constellations of the southern sky (many of which were named only within the last 250 years) are mythologically impoverished. In the North we have the Northern Cross, which is composed of all five stars that a cross deserves. It forms a subset of the larger constellation Cygnus the swan, which is flying across the sky along the Milky Way. Cygnus is nearly twelve times larger than the Southern Cross.

When people believe a tale that conflicts with self-checkable evidence it tells me that people undervalue the role of evidence in formulating an internal belief system. Why this is so is not clear, but it enables many people to hold fast to ideas and notions based purely on supposition. But all hope is not lost. Occasionally, people say things that are simply true no matter what. One of my favorites is, "Wherever you go, there you are" and its Zen corollary, "If we are all here, then we must not be all there."

THIRTY-FIVE

FEAR OF NUMBERS

We may never know the circuit diagram for all the electrochemical pathways within the human brain. But one thing is for certain, we are not wired for logical thinking. If we were, then mathematics would be the average person's easiest subject in school.

In this alternate universe, mathematics might not be taught at all because its foundations and principles would be self-evident even to slow-achieving students. But nowhere in the real world is this true. You can, of course, train most humans to be logical some of the time, and some humans to be logical all of the time; the brain is a marvellously flexible organ in this regard. But people hardly ever need training to be emotional. We are born crying, and we laugh early in life.

We do not emerge from the womb enumerating objects around us. The familiar number line, for example, is not writ on our gray matter. People had to invent the number line and build upon it when new needs arose from the growing complexities of life and of society. In a world of countable objects, we will all agree that $2 + 3 = 5$, but what does $2 - 3$ equal? To answer this question without saying, "It has no meaning," required that somebody invent a new part of the number line—negative numbers. Continuing: We all know that half of 10 is 5, but what is half of 5? To give meaning to this question, somebody had to invent fractions, yet another class of numbers on the number line. As this ascent through numberdom progressed, many more kinds of numbers would be invented: imaginary, irrational, transcendental, and complex, to name a few. They each have specific and sometimes unique applications to the physical world that we have discovered around us from the dawn of civilization.

Those who study the universe have been around from the beginning. As a member of this (second) oldest profession, I can attest that we have adopted, and actively use, all parts of the number line for all manner of heavenly analysis. We also routinely invoke some of the smallest and, of course, largest numbers of any profession. This state of mind has even influenced common parlance. When something in society seems immeasurably large, like the national debt, it's not called biological or chemical. It's called astronomical. And so one could argue strongly that astrophysicists do not fear numbers.

With thousands of years of culture behind us, what has society earned on its math report card? More specifically, what grade do we give Americans, members of the most technologically advanced culture the world has ever known?

Let's start with airplanes. Whoever lays out the seats on Continental Airlines seems to suffer from Medieval fears of the number 13. I have yet to see a row 13 on any flight I have taken with them. The rows simply go from 12 to 14. How about buildings? Seventy percent of all high-rises along a three-mile stretch of Broadway in Manhattan have no thirteenth floor. While I have not compiled detailed statistics for everywhere else in the nation, my experience walking in and out of buildings tells me it's more than half. If you've ridden the elevator of these guilty high-rises you've probably noticed that the 14th floor directly follows the 12th. This trend exists for old buildings as well as new. Some buildings are self-conscious and try to conceal their superstitious ways by providing two separate elevator banks: one that goes from 1 to 12 and another that goes upward from 14. The 22-story apartment building in which I was raised (in the Bronx) had two separate banks of elevators, but in this case, one bank accessed only the even floors while the other bank accessed the odd. One of the mysteries of my childhood was why the odd bank of elevators went from floor 11 directly to floor 15, and the even bank went from 12 to 16. Apparently, for my building, a single odd floor could not be skipped without throwing off the entire odd-even scheme. Hence the blatant omission of any reference to either the 13th *or* the 14th floor. Of course, all this meant that the building was actually only 20 stories high and not 22.

In another building, which harbored an extensive subterranean world, the levels below the first floor were B, SB, P, LB, and LL. Perhaps this is done to give you something to think about while you are otherwise standing in the elevator doing nothing. These floors are begging to become negative numbers. For the uninitiated, these abbreviations stood for Basement, Sub-Basement, Parking, Lower Basement, and Lower Level. We surely do not use such lingo to name normal floors. Imagine a building not with floors labelled 1, 2, 3, 4, 5, but G, AG, HG, VHG, SR, R, which obviously stand for Ground, Above Ground, High Ground, Very High Ground, Sub-Roof, and Roof. In principle, one should not fear negative floors—they don't in the Hotel de Rhone in Geneva, Switzerland, which has floors -1 and -2, nor are they afraid at the National Hotel in Moscow, which had no hesitation naming floors 0 and -1.

America's implicit denial of all that is less than zero shows up in many places. A mild case of this syndrome exists among car dealers, where instead of saying they will subtract $1,000 from the price of your car, they say you will receive $1,000 "cash back." In corporate accounting reports, we find that fear of the negative sign is pervasive. Here, it's common practice to enclose negative numbers in parentheses and not to display the negative symbol anywhere on the spreadsheet. Even the successful 1985 Bret Easton Ellis book (and 1987 film) *Less Than Zero*, which tracks the falling from grace of wealthy Los Angeles teens, could not be imagined with the logically equivalent title: *Negative*.

As we hide from negative numbers, we also hide from decimals, especially in America. Only recently have the stocks traded on the New York Stock Exchange been registered in decimal dollars instead of clunky fractions. And even though American money is decimal metric, we don't think of it that way. If something costs $1.50, we typically parse it into two segments and recite "one dollar and fifty cents." This behavior is not fundamentally different from the way people recited prices in the old decimal-averse British system that combined pounds and shillings.

When my daughter turned 15 months old, I took perverse pleasure in telling people she was "1.25." They would look back at me, with heads tilted in silent puzzlement, the way dogs look when they hear a high-pitched sound.

Fear of decimals is also rampant when probabilities are communicated to the public. People typically report odds in the form of "something to 1." Which makes intuitive sense to nearly everyone: The odds against the long-shot winning the ninth race at Belmont are 28 to 1. The odds against the favorite are 2 to 1. But the odds against the second favorite horse are 7 to 2. Why don't they say "something to 1"? Because if they did, then the 7 to 2 odds would instead read 3.5 to 1, stupefying all decimal-challenged people at the racetrack.

I suppose I can live with missing decimals, missing floors to tall buildings, and floors that are named instead of numbered. A more serious problem is the limited capacity of the human mind to grasp the relative magnitudes of large numbers:

Counting at the rate of one number per second, you will require 12 days to reach a million and 32 years to count to a billion. To count to a trillion takes 32,000 years, which is as much time as has elapsed since people first drew on cave walls.

If laid end to end, the hundred billion (or so) hamburgers sold by the McDonald's restaurant chain would stretch around the Earth 230 times leaving enough left over to stack the rest from Earth to the Moon—and back.

Last I checked, Bill Gates was worth $50 billion. If the average employed adult, who is walking in a hurry, will pick up a quarter from the sidewalk, but not a dime, then the corresponding amount of money (given their relative wealth) that Bill Gates would ignore if he saw it lying on the street is $25,000.

These are trivial brain exercises to the astrophysicist, but normal people do not think about these sorts of things. But at what cost? Beginning in 1969, space probes were designed and launched that shaped two decades of planetary reconnaissance in our solar system. The celebrated *Pioneer*, *Voyager*, and *Viking* missions were part of this era. So too was the *Mars Observer*, which was lost on arrival in the Martian atmosphere in 1993.

Each of these spacecraft took many years to plan and build. Each mission was ambitious in the breadth and depth of its scientific objectives and typically cost taxpayers between $1 and $2 billion. During a 1990s change in administration, NASA introduced a "faster, cheaper, better" paradigm for a new class of spacecraft that cost between $100 and $200 million. Unlike previous spacecraft, these could be planned and designed swiftly, enabling missions with more sharply defined objectives. Of course that meant a mission failure would be less costly and less damaging to the overall program of exploration.

In 1999, however, two of these more economical Mars missions failed, with a total hit to taxpayers of about $250 million. Yet public reaction was just as negative as it had been to the billion-dollar *Mars Observer*. The news media reported the $250 million as an unthinkably huge waste of money and proclaimed that something was wrong with NASA. The result was an investigation and a congressional hearing.

Not to defend failure, but $250 million is not much more than the cost to produce Kevin Costner's film flop *Waterworld*. It's also the cost of about two days in orbit for the space shuttle, and it's one-fifth the cost of the previously lost *Mars Observer*. Without these comparisons, and without the reminder that these failures were consistent with the "faster, cheaper, better" paradigm, in which risks are spread among multiple missions, you would think that $1 million equals $1 billion equals $1 trillion.

Nobody announced that the $250-million loss amounts to less than $1 per person in the United States. This much money, in the form of pennies, is surely just laying around in our streets, which are filled with people too busy to bend down and pick them up.

THIRTY-SIX

ON BEING BAFFLED

aybe it's the need to attract and keep readers. Maybe the public likes to know those rare occasions when scientists are clueless. But how come science writers can't write an article about the universe unless they describe some of the astrophysicists they interview as being "baffled" by the latest research headlines?

Scientific bafflement so intrigues journalists that, in what may have been a first for media coverage of science, an August 1999 page-one story in *The New York Times* reported on an object in the universe whose spectrum was a mystery (Wilford 1999). Top astrophysicists were stumped. In spite of the data's high quality (observations were made at the Hawaii-based Keck telescope, the most powerful optical observatory in the world), the object wasn't any known variety of planet, star, or galaxy. Imagine if a biologist had sequenced the genome of a newly discovered species of life and still couldn't classify it as plant or animal. Because of this fundamental ignorance, the 2,000-word article contained no analysis, no conclusions, no science.

In this particular case, the object was eventually identified as an odd, though otherwise unremarkable, galaxy—but not before millions of readers had been exposed to a parade of selected astrophysicists saying, "I dunno what it is." Such reporting is rampant, and grossly misrepresents our prevailing states of mind. If the writers told the whole truth, they would instead report that *all* astrophysicists are baffled *daily*, whether or not their research makes headlines.

Scientists cannot claim to be on the research frontier unless one thing or another baffles them. Bafflement drives discovery.

Richard Feynman, the celebrated twentieth-century physicist, humbly observed that figuring out the laws of physics is like observing a chess game without knowing the rules in advance. Worse yet, he wrote, you don't get to see each move in sequence. You only get to peek at the game in progress every now and then. With this intellectual handicap, your task is to deduce the rules of chess. You may eventually notice that bishops stay on a single color. That pawns don't move very fast. Or that a queen is feared by other pieces. But how about late in the game when only a few pawns are left. Suppose you come back and find one of the pawns missing and a previously captured queen resurrected in its place. Try to figure that one out. Most scientists would agree that the rules of the universe, whatever they may look like in their entirety, are vastly more complex than the rules of chess, and they remain a wellspring of endless bafflement.

I LEARNED RECENTLY that not all scientists are as baffled as astrophysicists. This could mean that astrophysicists are stupider than other breeds of scientists, but I think few would seriously make this claim. I believe that astrophysical bafflement flows from the staggering size and complexity of the cosmos. By this measure, astrophysicists have much in common with neurologists. Any one of them will assert, without hesitation, that what they do not know about the human mind vastly surpasses what they do know. That's why so many popular-level books are published annually on the universe and on the human consciousness—nobody's got it right yet. One might also include meteorologists in the ignorance club. So much goes on in Earth's atmosphere that can affect the weather, it's a wonder meteorologists predict anything accurately. The weather people on the evening news are the only reporters on the program who are expected to predict the news. They try hard to get it right but, in the end, all they can do is quantify their bafflement with statements like "50 percent chance of rain."

One thing is for certain, the more profoundly baffled you have been in your life, the more open your mind becomes to new ideas. I have firsthand evidence of this.

During an appearance on the PBS talk show *Charlie Rose*, I was pitted against a well-known biologist to discuss and evaluate the evidence for extraterrestrial life as revealed in the nooks and crannies of the now-famous Martian meteorite ALH84001. This potato-shaped, potato-sized interplanetary traveler was thrust off the Martian surface by the impact of an energetic meteor, in a manner not unlike what happens to loose Cheerios as they get thrust off a bed when you jump up and down on the mattress. The Martian meteorite then traveled through interplanetary space for tens of millions of years, crashed into Antarctica, stayed buried in ice for about 10,000 years, and was finally recovered in 1984.

The original 1996 research paper by David McKay and colleagues presented a string of circumstantial evidence. Each item, by itself, could be ascribed to a nonbiogenic process. But taken together, they made a compelling case for Mars's having once harbored life. One of McKay's most intriguing, but scientifically empty, pieces of evidence was a simple photograph of the rock, taken with a high-resolution microscope showing a teeny-weeny worm-looking thing, less than one-tenth the size of the smallest known worm creatures on Earth. I was (and still am) quite enthusiastic about these findings. But my biology co-panelist was argumentatively skeptical. After he chanted Carl Sagan's mantra "extraordinary claims require extraordinary evidence" a few times, he declared that the wormy thing could not possibly be life because there was no evidence of a cell wall and that it was much smaller than the smallest known life on Earth.

Excuse me?

Last I checked the conversation was about Martian life, not the

Earth life he had grown accustomed to studying in his laboratory. I could not imagine a more close-minded statement. Was I being irresponsibly open-minded? It is, indeed, possible to be so open-minded that important mental faculties have spilled out, like those who are prone to believe, without skepticism, reports of flying saucers and alien abductions. How is it that my brain could be wired so differently from that of the biologist? He and I both went to college, then graduate school. We got our PhDs in our respective fields and have devoted our lives to the methods and tools of science. Perhaps we needn't look far for the answer. Publicly and among themselves biologists rightly celebrate the diversity of life on Earth from the marvelous variations wrought by natural selection and expressed by differences in DNA from one species to the next. At the end of the day, however, their confession is heard by no one: they work with a single scientific sample—life on Earth.

I'D BET ALMOST anything that life from another planet, if formed independently from life on Earth, would be more different from all species of Earth life than any two species of Earth life are from each other. On the other hand, the objects, classification schemes, and data sets of the astrophysicist are drawn from the entire universe. For this simple reason, new data routinely pushes astrophysicists to think outside the proverbial box. And sometimes our whole bodies get shoved completely outside the box.

We could go back to ancient times for examples, but that's unnecessary. The twentieth century will do just fine. And many of these examples we have already discussed:

Just when we thought it was safe to look up at a clockwork universe, and bask in our deterministic laws of classical physics, Max Planck, Werner Heisenberg, and others had to go and discover quantum mechanics, demonstrating that the smallest scales of the universe are inherently nondeterministic even if the rest of it is.

Just when we thought it was safe to talk about the stars of the night sky as the extent of the known cosmos, Edwin Hubble had to go and discover that all the spiral fuzzy things in the sky were external galaxies. Veritable "island universes," adrift far beyond the extent of the Milky Way's stars.

Just when we thought we had the size and shape of our presumably eternal cosmos figured out, Edwin Hubble went on to discover that the universe was expanding and that the galactic universe extended as far as the largest telescopes could see. One consequence of this discovery was that the cosmos had a beginning—an unthinkable notion to all previous generations of scientists.

Just when we thought that Albert Einstein's relativity theories would enable us to explain all the gravity of the universe, the Caltech astrophysicist Fritz Zwicky discovered dark matter, a mysterious substance that wields 90 percent of all the gravity of the universe, but emits no light and has no other interactions with ordinary matter. The stuff is still a mystery. Fritz Zwicky further identifies and characterizes a class of objects in the universe called supernovas, which are single, exploding stars that temporarily emit the energy equivalent of a hundred billion suns.

Not long after we figured out the ways and means of supernova explosions, somebody discovered bursts of gamma rays from the edge of the universe that temporarily outshined all the energy-emitting objects of the rest of the universe combined.

And just as we were growing accustomed to living in our ignorance of dark matter's true nature, two research groups working independently, one led by Berkeley astrophysicist Saul Perlmutter and one led by Harvard astrophysicist Robert Kirshner, discovered that the universe is not just expanding, it's accelerating. The cause? Evidence indicates a mysterious pressure within the vacuum of space that acts in the opposite direction of gravity and which remains more of a mystery than dark matter.

These are, of course, just an assortment of the countless mind-bending and brain-boggling phenomena that have kept astrophysicists busy for the past hundred years. I could stop the list here, but I would be remiss if I did not include the discovery of neutron stars, which pack the mass of the Sun within a ball that measures barely a dozen miles across. To achieve this density at home, just cram a herd of 50 million elephants into the volume of a thimble.

No doubt about it. My mind is wired differently from that of a biologist, and so our different reactions to the evidence for life in the Mars meteorite was understandable, if not entirely expected.

Lest I leave you with the impression that the behavior of research scientists is indistinguishable from that of freshly beheaded chickens running aimlessly around the coup, you should know that the body of knowledge about which scientists are not baffled is impressive. It forms most of the contents of introductory college textbooks and comprises the modern consensus of how the world works. These ideas are so well understood that they no longer form interesting subjects of research and are no longer a source of confusion.

I once hosted and moderated a panel discussion on theories of everything—those wishful attempts to explain under one conceptual umbrella all the forces of nature. On the stage were five distinguished and well-known physicists. Midway through the debate I nearly had to break up a fight as one of them looked like he was ready to throw a punch. That's okay. I didn't mind it. The lesson here is if you ever see scientists engaged in a heated debate, they are arguing because they are all baffled. These physicists were arguing on the frontier about the merits and shortcomings of string theory, not whether Earth orbits the Sun, or whether the heart pumps blood to the brain, or whether rain falls from clouds.

THIRTY-SEVEN

FOOTPRINTS IN THE
SANDS OF SCIENCE

I f you visit the gift shop at the Hayden Planetarium in New York City, you'll find all manner of space-related paraphernalia for sale. Familiar things are there—plastic models of the space shuttle and the *International Space Station*, cosmic refrigerator magnets, Fisher space pens. But unusual things are there too—dehydrated astronaut ice cream, astronomy Monopoly, Saturn-shaped salt-and-pepper shakers. And that's not to mention the weird things such as *Hubble* telescope pencil erasers, Mars rock super-balls, and edible space worms. Of course, you'd expect a place like the planetarium to stock such stuff. But something much deeper is going on. The gift shop bears silent witness to the iconography of a half-century of American scientific discovery.

In the twentieth century, astrophysicists in the United States discovered galaxies, the expanding of the universe, the nature of supernovas, quasars, black holes, gamma-ray bursts, the origin of the elements, the cosmic microwave background, and most of the known planets in orbit around solar systems other than our own. Although the Russians reached one or two places before us, we sent space probes to Mercury, Venus, Jupiter, Saturn, Uranus, and Neptune. American probes have also landed on Mars and on the asteroid Eros. And American astronauts have walked on the Moon. Nowadays most Americans take all this for granted, which is practically a working definition of culture: something everyone does or knows about, but no longer actively notices.

While shopping at the supermarket, most Americans aren't surprised to find an entire aisle filled with sugar-loaded, ready-to-eat breakfast cereals. But foreigners notice this kind of thing immediately, just as traveling Americans notice that supermarkets in Italy display vast selections of pasta and that markets in China and Japan offer an astonishing variety of rice. The flip side of not noticing your own culture is one of the great pleasures of foreign travel: realizing what you hadn't noticed about your own country, and noticing what the people of other countries no longer realize about themselves.

Snobby people from other countries like to make fun of the U.S. for its abbreviated history and its uncouth culture, particularly compared with the millennial legacies of Europe, Africa, and Asia. But 500 years from now historians will surely see the twentieth century as the American century—the one in which American discoveries in science and technology rank high among the world's list of treasured achievements.

Obviously the U.S. has not always sat atop the ladder of science. And there's no guarantee or even likelihood that American preeminence will continue. As the capitals of science and technology move from one nation to another, rising in one era and falling in the next, each culture leaves its mark on the continual attempt of our species to understand the universe and our place in it. When historians write their accounts of such world events, the traces of a nation's presence on center stage sit prominently in the timeline of civilization.

MANY FACTORS INFLUENCE how and why a nation will make its mark. Strong leadership matters. So does access to resources. But something else must be present—something less tangible, but with the power to drive an entire nation to focus its emotional, cultural, and intellectual capital on creating islands of excellence in the world. Those who live in such times often take for granted what they have created, on the blind assumption that things will continue forever as they are, leaving their achievements susceptible to abandonment by the very culture that created it.

Beginning in the 700s and continuing for nearly 400 years—while Europe's Christian zealots were disemboweling heretics—the Abbasid caliphs created a thriving intellectual center of arts, sciences, and medicine for the Islamic world in the city of Baghdad. Muslim astronomers and mathematicians built observatories, designed advanced timekeeping tools, and developed new methods of mathematical analysis and computation. They preserved the extant works of science from ancient Greece and elsewhere and translated them into Arabic. They collaborated with Christian and Jewish scholars. And Baghdad became a center of enlightenment. Arabic was, for a time, the lingua franca of science.

The influence of these early Islamic contributions to science remains to this day. For example, so widely distributed was the Arabic translation of Ptolemy's magnum opus on the geocentric universe (originally written in Greek in A.D. 150), that even today, in all translations, the work is known by its Arabic title *Almagest,* or "The Greatest."

The Iraqi mathematician and astronomer Muhammad ibn Musa al-Khwarizmi gave us the words "algorithm" (from his name, al-Khwarizmi) and "algebra" (from the word *al-jabr* in the title of his book on algebraic calculation). And the world's shared system of numerals—0, 1, 2, 3, 4, 5, 6, 7, 8, 9—though Hindi in origin, were neither common nor widespread until Muslim mathematicians exploited them. The Muslims furthermore made full and innovative use of the zero, which did not exist among Roman numerals or in any established numeric system. Today, with legitimate reason, the ten symbols are internationally referred to as Arabic numerals.

PORTABLE, ORNATELY ETCHED, brass astrolabes were also developed by Muslims, from ancient prototypes, and became as much works of art as tools of astronomy. An astrolabe projects the domed heavens onto a flat surface and, with layers of rotating and nonrotating dials, resembles the busy, ornate face of a grandfather clock. It enabled astronomers, as well as others, to measure the positions of the Moon and the stars on the sky, from which they could deduce the time—a generally useful thing to do, especially when it's time to pray. The astrolabe was so popular and influential as a terrestrial connection to the cosmos that, to this day, nearly two-thirds of the brightest stars in the night sky retain their Arabic names.

The name typically translates into an anatomical part of the constellation being described. Famous ones on the list (along with their loose translations) include: Rigel (*Al Rijl,* "foot") and Betelgeuse (*Yad al Jauza,* "hand of the great one"—in modern times drawn as the armpit), the two brightest stars in the constellation Orion; Altair (*At-Ta'ir,* "the flying one"), the brightest star in the constellation Aquila, the eagle; and the variable star Algol (*Al-Ghul,* "the ghoul"), the second brightest star in the constellation Perseus, referring to the blinking eye of the bloody severed head of Medusa held aloft by Perseus. In the less-famous category are the two brightest stars of the constellation Libra, athough identified with the scorpion in the heyday of the astrolabe: Zubenelgenubi (*Az-Zuban al-Janubi,* "southern claw") and Zebueneschamali (*Az-Zuban ash-Shamali,* "northern claw"), the longest surviving star names in the sky.

At no time since the eleventh century has the scientific influence of the Islamic world been equal to what it enjoyed the preceding four centuries. The late Pakistani physicist Abdus Salam, the first Muslim ever to win the Nobel Prize, lamented:

> *There is no question [that] of all civilizations on this planet, science is the weakest in the lands of Islam. The dangers of this weakness cannot be overemphasized since honorable survival of a society depends directly on strength in science and technology in the conditions of the present age.* (Hassan and Lui 1984, p. 231)

PLENTY OF OTHER nations have enjoyed periods of scientific fertility. Think of Great Britain and the basis of Earth's system of longitude. The prime meridian is the line that separates geographic east from west on the globe. Defined as zero degrees longitude, it bisects the base of a telescope at an observatory in Greenwich, a London borough on the south bank of the River Thames. The line doesn't pass through New York City. Or Moscow. Or Beijing. Greenwich was chosen in 1884 by an international consortium of longitude mavens who met in Washington, DC, for that very purpose.

By the late nineteenth century, astronomers at the Royal Greenwich Observatory—founded in 1675 and based, of course, in Greenwich—had accumulated and catalogued a century's worth of data on the exact positions of thousands of stars. The Greenwich astronomers used a common but specially designed telescope, constrained to move along the meridional arc that connects due north to due south through the observer's zenith. By not tracking the general east to west motion of the stars, they simply drift by as Earth rotates. Formally known as a transit instrument, such a telescope allows you to mark the exact time a star crosses your field of view. Why? A star's "longitude" on the sky is the time on a sidereal clock the moment the star crosses your meridian. Today we calibrate our watches with atomic clocks, but back then there was no timepiece more reliable than the rotating Earth itself. And there was no better record of the rotating Earth than the stars that passed slowly overhead. And nobody measured the positions of passing stars better than the astronomers at the Royal Greenwich Observatory.

During the seventeenth century Great Britain had lost many ships at sea due to the challenges of navigation that result from not knowing your longitude with precision. In an especially tragic disaster in 1707, the British fleet, under Vice Admiral Sir Clowdesley Shovell, ran aground into the Scilly Isles, west of Cornwall, losing four ships and 2,000 men. With enough impetus, England finally commissioned a Board of Longitude, which offered a fat cash award—£20,000—to the first person who could design an ocean-worthy chronometer. Such a timepiece was destined to be important in both military and commercial ventures. When synchronized with the time at Greenwich, such a chronometer could determine a ship's longitude with great precision. Just subtract your local time (readily obtained from the observed position of the Sun or stars) from the chronometer's time. The difference between the two is a direct measure of your longitude east or west of the prime meridian.

In 1735, the Board of Longitude's challenge was met by a portable, palm-sized clock designed and built by an English mechanic, John Harrison. Declared to be as valuable to the navigator as a live person standing watch at a ship's bow, Harrison's chronometer gave renewed meaning to the word "watch."

Because of England's sustained support for achievements in astronomical and navigational measurements, Greenwich landed the prime meridian. This decree fortuitously placed the international date line (180 degrees away from the prime meridian) in the middle of nowhere, on the other side of the globe in the Pacific Ocean. No country would be split into two days, leaving it beside itself on the calendar.

IF THE ENGLISH have forever left their mark on the spatial coordinates of the globe, our basic temporal coordinate system—a solar-based calendar—is the product of an investment of science within the Roman Catholic Church. The incentive to do so was not driven by cosmic discovery itself but by the need to keep the date for Easter in the early spring. So important was this need that Pope Gregory XIII established the Vatican Observatory, staffing it with erudite Jesuit priests who tracked and measured the passage of time with unprecedented accuracy. By decree, the date for Easter had been set to the first Sunday after the first full moon after the vernal equinox (preventing Holy Thursday, Good Friday, and Easter Sunday from ever falling on a special day in somebody else's lunar-based calendar). That rule works as long as the first day of spring stays in March, where it belongs. But the Julian calendar of Julius

Caesar's Rome was sufficiently inaccurate that by the sixteenth century it had accumulated 10 extra days, placing the first day of spring on April 1 instead of March 21. The four-year leap day, a principal feature of the Julian calendar, had slowly overcorrected the time, pushing Easter later and later in the year.

In 1582, when all the studies and analyses were complete, Pope Gregory deleted the 10 offending days from the Julian calendar and decreed the day after October 4 to be October 15. The Church thenceforth made an adjustment: for every century year not evenly divisible by 400, a leap day gets omitted that would otherwise have been counted, thus correcting for the overcorrecting leap day itself.

This new "Gregorian" calendar was further refined in the twentieth century to become even more precise, preserving the accuracy of your wall calendar for tens of thousands of years to come. Nobody else had ever kept time with such precision. Enemy states of the Catholic Church (such as Protestant England, and its rebellious progeny, the American colonies) were slow to adopt the change, but eventually everyone in the civilized world, including cultures that traditionally relied on Moon-based calendars, adopted the Gregorian calendar as the standard for international business, commerce, and politics.

EVER SINCE THE BIRTH of the industrial revolution the European contributions to science and technology have become so embedded in Western culture that it may take a special effort to step outside and notice them at all. The revolution was a breakthrough in our understanding of energy, enabling engineers to dream up ways to convert it from one form to another. In the end, the revolution would serve to replace human power with machine power, drastically enhancing the productivity of nations and the subsequent distribution of wealth around the world.

The language of energy is rich with the names of those scientists who contributed to the effort. James Watt, the Scottish engineer who perfected the steam engine in 1765, has the moniker best known outside the circles of engineering and science. Either his last name or his monogram gets stamped on the top of practically every lightbulb. A bulb's wattage measures the rate it consumes energy, which correlates with its brightness. Watt worked on steam engines while at the University of Glasgow, which was, at the time, one of the world's most fertile centers for engineering innovation.

The English physicist Michael Faraday discovered electromagnetic induction in 1831, which enabled the first electric motor. The farad, a measure of a device's capacity to store electric charge, probably doesn't do full justice to his contributions to science.

The German physicist Heinrich Hertz discovered electromagnetic waves in 1888, which enabled communication via radio; his name survives as the unit of frequency along with its metric derivatives "kilohertz," "megahertz," and "gigahertz."

From the Italian physicist Alessandro Volta we have the volt, a unit of electric potential. From the French physicist André-Marie Ampère, we have the unit of electric current known as the ampere, or "amp" for short. From the British physicist James Prescott Joule, we have the joule, a unit of energy. The list goes on and on.

With the exception of Benjamin Franklin and his tireless experiments with electricity, the U.S. as a nation watched this fertile chapter of human achievement from afar, preoccupied with gaining its independence from England and exploiting the economies of slave labor. Today the best we could do was pay homage in the original *Star Trek* television series: Scotland is the country of origin of the industrial revolution, and of the chief engineer of the starship *Enterprise*. His name? "Scotty" of course.

In the late eighteenth century the industrial revolution was in full swing, but so too was the French Revolution. The French used the occasion to shake up more than the royalty; they also introduced the metric system to standardize what was then a world of mismatched measures—confounding science and commerce alike. Members of the French Academy of Sciences led the world in measures of the Earth's shape and had proudly determined it to be an oblate spheroid. Building on this knowledge, they defined the meter to be one ten-millionth the distance along the Earth's surface from the North Pole to the equator, passing through—where else?— Paris. This measure of length was standardized as the separation between two marks etched on a special bar of platinum alloyed with iridium. The French devised many other decimal standards that (except for decimal time and decimal angles) were ultimately adopted by all the civilized nations of the world except the U.S., the west African nation of Liberia, and the politically unstable tropical nation of Myanmar. The original artifacts of this metric effort are preserved at the International Bureau of Weights and Measures— located, of course, near Paris.

BEGINNING IN THE late 1930s the U.S. became a nexus of activity in nuclear physics. Much of the intellectual capital grew out of the exodus of scientists from Nazi Germany. But the financial capital came from Washington, in the race to beat Hitler to build an atomic bomb. The coordinated effort to produce the bomb was known as the Manhattan Project, so named because much of the early research had been done in Manhattan, at Columbia University's Pupin Laboratories.

The wartime investments had huge peacetime benefits for the community of nuclear physicists. From the 1930s through the 1980s, American accelerators were the largest and most productive in the world. These racetracks of physics are windows into the fundamental structure and behavior of matter. They create beams of subatomic particles, accelerate them to near the speed of light with a cleverly configured electric field, and smash them into other particles, busting them to smithereens. Sorting through the smithereens, physicists have found evidence for hoards of new particles and even new laws of physics.

American nuclear physics labs are duly famous. Even people who are physics-challenged will recognize the top names: Los Alamos; Lawrence Livermore; Brookhaven; Lawrence Berkeley; Fermi Labs; Oak Ridge. Physicists at these places discovered new particles, isolated new elements, informed a nascent theoretical model of particle physics, and collected Nobel Prizes for doing so.

The American footprint in that era of physics is forever inscribed at the upper end of the periodic table. Element number 95 is americium; number 97 is berkelium; number 98 is californium; number 103 is lawrencium, for Ernest O. Lawrence, the American physicist who invented the first particle accelerator; and number 106 is seaborgium, for Glenn T. Seaborg, the American physicist whose lab at the University of California, Berkeley, discovered ten new elements heavier than uranium.

EVER-LARGER ACCELERATORS reach ever-higher energies, probing the fast-receding boundary between what is known and unknown about the universe. The big bang theory of cosmology asserts that the universe was once a very small and very hot soup of energetic subatomic particles. With a super-duper particle-smasher, physicists might be able to simulate the earliest moments of the cosmos. In the 1980s, when U.S. physicists proposed just such an accelerator (eventually dubbed the Super-conducting Super Collider), Congress was ready to fund it. The U.S. Department of Energy was ready to oversee it. Plans were drawn up. Construction began. A circular tunnel 50 miles around (the size of the Washington, DC, beltway) was dug in Texas. Physicists were eager to peer across the next cosmic frontier. But in 1993, when cost overruns looked intractable, a fiscally frustrated Congress permanently withdrew funds for the $11 billion project. It probably never occurred to our elected representatives that by canceling the Super Collider they surrendered America's primacy in experimental particle physics.

If you want to see the next frontier, hop a plane to Europe, which seized the opportunity to build the world's largest particle accelerator and stake a claim of its own on the landscape of cosmic knowledge. Known as the Large Hadron Collider, the accelerator will be run by the European Center for Particle Physics (better known by an acronym that no longer fits its name, CERN). Although some U.S physicists are collaborators, America as a nation will watch the effort from afar, just as so many nations have done before.

THIRTY-EIGHT

LET THERE BE DARK

Astrophysics reigns as the most humbling of scientific disciplines. The astounding breadth and depth of the universe deflates our egos daily, and we are continually at the mercy of uncontrolled forces. A simple cloudy evening—one that would stop no other human activity—prevents us from making observations with a telescope that can cost $20,000 a night to run, regardless of the weather. We are passive observers of the cosmos, acquiring data when, where, and how nature reveals it to us. To know the cosmos requires that we have windows onto the universe that remain unfogged, untinted, and unpolluted. But the spread of what we call civilization, and the associated ubiquity of modern technology, is generally at odds with this mission. Unless we do something about it, people will soon bathe Earth in a background glow of light, blocking all access to the frontiers of cosmic discovery.

The most obvious and prevalent form of astropollution comes from streetlamps. All too often, they can be seen from your airplane window during night flights, which means that these streetlamps illuminate not only the streets below but the rest of the universe. Unshielded streetlights, such as those without downward-facing shades, are most to blame. Municipalities with these poorly designed lamp housings find themselves buying higher-wattage bulbs because half the lamplight points upward. This wasted light, shot forth into the night sky, has rendered much of the world's real estate unsuitable for astronomical research. At the 1999 "Preserving the Astronomical Sky" symposium, participants rightly moaned about the loss of dark skies around the globe. One paper reported that inefficient lighting costs the city of Vienna $720,000 annually; London $2.9 million; Washington, DC, $4.2 million; and New York City $13.6 million (Sullivan and Cohen 1999, pp. 363–68). Note that London, with a population similar to that of New York City, is more efficient in its inefficiency by nearly a factor of 5.

The astrophysicist's quandary is not that light escapes into space but that the lower atmosphere supports a mixture of water vapor, dust, and pollutants that bounce some of the upward-flowing photons back down to Earth, leaving the sky aglow with the signature of a city's nightlife. As cities become brighter and brighter, dim objects in the cosmos become less and less visible, severing urban dwellers' access to the universe.

It's hard to exaggerate the magnitude of this effect. A penlight's beam, aimed at a wall across a darkened dining room, is easy to spot. But gradually brighten the overhead light, and watch how the beam gets harder and harder to see. Under light-polluted skies, fuzzy objects such as comets, nebulae, and galaxies become difficult or impossible to detect. I have never in my life seen the Milky Way galaxy from within the limits of New York City, and I was born and raised here. If you observe the night sky from light-drenched Times Square, you might see a dozen or so stars, compared with the thousands that were visible from the same spot when Peter Stuyvesant was hobbling around town. No wonder ancient peoples shared a culture of sky lore, whereas modern peoples, who know nothing of the night sky, instead share a culture of evening TV.

The expansion of electrically lit cities during the twentieth century created a technology fog that forced astronomers to move their hilltop observatories from the outskirts of towns to remote places such as the Canary Islands, the Chilean Andes, and Hawaii's Mauna Kea. One notable exception is Kitt Peak National Observatory in Arizona. Instead of running away from the spreading and brightening city of Tucson, 50 miles away, the astronomers stayed and fought. The battle is easier won than you might think; all you have to do is convince people that their choice of outdoor lighting is a waste of money. In the end, the city gets efficient streetlamps and the astronomers get a dark sky. Ordinance No. 8210 of the Tucson/Pima County Outdoor Lighting Code reads as though the mayor, the chief of police, and the prison warden were all astronomers at the time the code was passed. Section 1 identifies the intent of the ordinance:

> *The purpose of this Code is to provide standards for outdoor lighting so that its use does not unreasonably interfere with astronomical observations. It is the intent of this Code to encourage, through the regulation of the types, kinds, construction, installation, and uses of outdoor electrically powered illuminating devices, lighting practices and systems to conserve energy without decreasing safety, utility, security, and productivity while enhancing nighttime enjoyment of property within the jurisdiction.*

And after 13 other sections that give strict rules and regulations governing citizens' choice of outdoor lighting, we get to the best part, section 15:

> *It shall be a civil infraction for any person to violate any of the provisions of this Code. Each and every day during which the violation continues shall constitute a separate offense.*

As you can see, by shining light on an astronomer's telescope you can turn a peace-loving citizen into a Rambo. Think I'm joking? The International Dark-Sky Association (IDA) is an organization that fights upward-pointing light anywhere in the world. With an opening phrase reminiscent of the one painted on Los Angeles Police Department squad cars, the IDA's motto says it all: "To preserve and protect the nighttime environment and our heritage of dark skies through quality outdoor lighting." And, like the police, the IDA will come after you if you transgress.

I know. They came after me. Not a week after the Rose Center for Earth and Space first opened its doors to the public, I received a letter from the IDA's executive director, scolding me for the upward-pointing lights embedded in the pavement of our entrance plaza. We were justly accused—the plaza does have forty (very low wattage) lamps that help delineate and illuminate the Rose Center's granite-clad arched entryway. These lights are partly functional and partly decorative. The point of the letter was not to blame the bad viewing conditions across all of New York City on these itty-bitty lamps but to hold the Hayden Planetarium accountable for setting a good example for the rest of the world. I am embarrassed to say that the lights remain.

But all that's bad is not artificial. A full Moon is bright enough to reduce the number of stars visible to the unaided eye from thousands to hundreds. Indeed, the full Moon is more than 100,000 times brighter than the brightest nighttime stars. And the physics of reflection angles endows the full Moon with more than ten times the brightness of a half Moon. This moonglow also greatly reduces the number of meteors visible during a meteor shower (though clouds would be worse), no matter where you are on Earth. So never wish a full Moon upon an astronomer who is headed off to a big telescope. True, the Moon's tidal force created tide pools and other dynamic habitats that contributed to the transition from marine to terrestrial life and ultimately made it possible for humans to thrive. Apart from this detail, most observational astronomers, especially cosmologists, would be happy if the Moon had never existed.

A few years ago I got a phone call from a marketing executive who wanted to light up the Moon with the logo of her company. She wanted to know how she might proceed. After slamming down the phone, I called her back and politely explained why it was a bad idea. Other corporate executives have asked me how to put into orbit mile-wide luminous banners with catchy slogans written across them, much like the skywriting or flag-dragging airplanes you see at sports events or over the ocean from a crowded beach. I always threaten to send the light police after them.

Modern life's insidious link with light pollution extends to other parts of the electromagnetic spectrum. Next at risk is the astronomer's radio-wave window to the cosmos, including microwaves. In modern times we are awash in the signals of such radio wave–emitting devices as cellular telephones, garage-door openers, keys that trigger "boip" sounds as they remotely lock and unlock car doors, microwave relay stations, radio and television transmitters, walkie-talkies, police radar guns, Global Positioning Systems, and satellite communications networks. Earth's radio-wave window to the universe lies cloaked in this technologically induced fog. And the few clear bands that remain within the radio spectrum are getting progressively narrower as the trappings of high-tech living grab more and more radio-wave real estate. The detection and study of extremely faint celestial objects is being compromised as never before.

In the past half-century radio astronomers discovered remarkable things, including pulsars, quasars, molecules in space, and the cosmic microwave background, the first evidence in support of the big bang itself. But even a wireless conversation can drown such faint radio signals: modern radio telescopes are so sensitive that a cell-phone encounter between two astronauts on the Moon would be one of the brightest sources in the radio sky. And if Martians used cell phones, our most powerful radio telescopes would easily nab them, too.

The Federal Communications Commission is not unmindful of the heavy, often conflicting demands that various segments of society place on the radio spectrum. The FCC's Spectrum Policy Task Force intends to review the policies that govern use of the electromagnetic spectrum, with the goal of improving efficiency and flexibility. FCC chairman Michael K. Powell told the *Washington Post* (June 19, 2002) that he wanted the FCC's philosophy to shift from a "command and control" approach to a "market-oriented" one. The commission will also review how it allocates and assigns bands of the radio spectrum, as well as how one allocation may interfere with another.

For its part, the American Astronomical Society, the professional organization of the nation's astrophysicists, has called on its members to be as vigilant as the IDA folks—a posture I endorse—in trying to convince policy makers that specially identified radio frequencies should be left clear for astronomers' use. To borrow vocabulary and concepts from the irrepressible Green movement, these bands should be considered a kind of "electromagnetic wilderness" or "electromagnetic national park." To eliminate interference, the geographic areas surrounding the protected observatories should also be kept clear of human-generated radio signals of any kind.

The most challenging problem may be that the farther an object is from the Milky Way, the longer the wavelength and the lower the frequency of its radio signals. This phenomenon, which is a cosmological Doppler effect, is the principal signature of our expanding universe. So it's not really possible to isolate a single range of "astro" frequencies and assert that the entire cosmos, from nearby galaxies to the edge of the observable universe, can be served through this window. The struggle continues.

Today, the best place to build telescopes for exploring all parts of the electromagnetic spectrum is the Moon. But not on the side that faces Earth. Putting them there might be worse than looking out from Earth's surface. When viewed from the Moon's near side, Earth looks thirteen times bigger, and shines some fifty times brighter, than the Moon does when viewed from Earth. And Earth never sets. As you might suspect, civilization's chattering communication signals also make Earth the brightest object in the radio-wave sky. The astronomer's heaven is, instead, the Moon's far side, where Earth never rises, remaining forever buried below the horizon.

Without a view of Earth, telescopes built on the Moon could point in any skyward direction, without the risk of contamination from Earth's electromagnetic emanations. Not only that, night on the Moon lasts nearly 15 Earth days, which would enable astronomers to monitor objects in the sky for days on end, much longer than they can from Earth. And because there is no lunar atmosphere, observations conducted from the Moon's surface would be as good as observations of the cosmos from Earth orbit. The *Hubble Space Telescope* would lose the bragging rights it now enjoys.

Furthermore, without an atmosphere to scatter sunlight, the Moon's daytime sky is almost as dark as its night, so everybody's favorite stars hover visibly in the sky, right alongside the disk of the Sun. A more pollution-free place has yet to be found.

On second thought, I retract my earlier callous remarks about the Moon. Maybe our neighbor in space will one day become the astronomer's best friend after all.

HOLLYWOOD NIGHTS

Few things are more annoying to avid moviegoers than being accompanied to a film by hyperliterate friends who can't resist making comments about why the book was better. These people babble on about how the characters in the novel were more fully developed or how the original story line was more deeply conceived. In my opinion, they should just stay home and leave the rest of us to enjoy the film. For me, it's purely a matter of economics: to see a movie is cheaper and faster than to buy and read the book on which it was based. With this anti-intellectual attitude, I ought to be mute every time I notice scientific transgressions in a movie's story or set design. But I am not. On occasion, 1 can be as annoying as bookworms to other moviegoers. Over the years, I have collected especially egregious errors in Hollywood's attempts to show or engage the cosmos. And I can no longer keep them to myself.

My list, by the way, does not consist of bloopers. A blooper is a mistake that the producers or continuity editors happen to miss, but normally catch and fix. The astro-errors I'm talking about were willingly introduced and indicate a profound lack of attention to easily checkable detail. I would further assert that none of these writers, producers, or directors ever took Astronomy 101 in college.

Let's start at the bottom.

At the end of the 1977 Disney film *Black Hole,* which sits on many people's 10 worst movies list (including mine), an H. G.

Wellsian spaceship loses control of its engines and plunges into a black hole. What more could special-effects artists ask for? Let's see how well they did. Was the craft and its crew ripped apart by the ever-increasing tidal forces of gravity—something a real black hole would do to them? No. Was there any attempt to portray relativistic time dilation, as predicted by Einstein, where the universe around the doomed crew evolves rapidly over billions of years while they, themselves, age only a few ticks of their wristwatches? No. The scene did portray a swirling disk of accreted gas around the black hole. Good. Black holes do this sort of thing with gas that falls toward them. But did elongated jets of matter and energy spew forth from each side of the accretion disk? No. Did the ship travel through the black hole and get spit out into another time? another part of the universe? or in another universe altogether? No. Instead of capturing these cinematically fertile and scientifically informed ideas, the storytellers depicted the black hole's innards as a dank cave, with fiery stalagmites and stalactites, as though we were touring Carlsbad Cavern's hot and smoky basement.

Some people may think of these scenes as expressions of the director's poetic or artistic license, allowing him to invent whimsical cosmic imagery without regard to the real universe. But given how lame the scenes were, they are more likely to have been an expression of the director's scientific ignorance. Suppose there were such a thing as "scientific license," where a scientist, doing art, chooses to ignore certain fundamentals of artistic expression. Suppose that whenever scientists drew a woman they gave her three breasts, seven toes on each foot, and an ear in the middle of her face? In a less extreme example, suppose scientists drew people with the knee joint bending the wrong way, or with odd proportions among the body's long bones? If this did not start a new movement in artistic expression—akin to Picasso's flounderlike renderings of the human face—then artists would surely tell us all to go back to school immediately and take some art classes in basic anatomy.

Was it license or ignorance that led the painter of a work in the Louvre to draw a cul-de-sac surrounded by erect trees, each with a

Sun-made shadow pointing in toward the center of the circle? Hadn't the artist ever noticed that all shadows cast by the Sun on vertical objects are parallel? Is it license or ignorance that nearly every Moon ever painted by artists is either a crescent or a full moon? Half of any month the Moon's phase is neither crescent nor full. Did the artists paint what they saw or what they wished they had seen? When Francis Ford Coppola's 1987 *Someone to Watch Over Me* was being filmed, his cinematographer called my office to ask when and where was the best occasion to film the full Moon rising over the Manhattan skyline. When I instead offered him the first quarter moon or the waxing gibbous moon, he was unimpressed. Only the full Moon would do.

Although I rant, there's no doubt that creative contributions from the world's artists would be poorer in the absence of artistic license. Among other losses, there would have been no impressionism or cubism. But what distinguishes good artistic license from bad is whether the artist acquired access to all relevant information before the creativity begins. Perhaps Mark Twain said it best:

Get your facts first, and then you can distort 'em as much as you please. (1899, Vol. 2, Chap. XXXVII)

In the 1997 blockbuster movie *Titanic*, producer and director James Cameron not only invested heavily in special effects but also in re-creating the ship's luxurious interiors. From the wall sconces to the patterns on the china and silverware, no detail of design was too small to attract the attention of Mr. Cameron, who made sure to reference the latest artifacts salvaged from missions to the sunken ship, more than two miles undersea. Furthermore, he carefully researched the history of fashion and social mores to ensure that his characters dressed and behaved in ways generally consistent with the year 1912. Aware that the ship was designed with only three of its four smoke stacks connected to engines, Cameron accurately portrays smoke coming from only three stacks. We know from accurate records of this maiden voyage from Southampton to New York City the date and time the ship sank, as well as the longitude and latitude on Earth where it sank. Cameron captures that too.

With all this attention to detail, you think James Cameron might have paid a bit more attention to which stars and constellations were visible on that fateful night.

He didn't.

In the movie, the stars above the ship bear no correspondence to any constellations in a real sky. Worse yet, while the heroine bobs and hums a tune on a slab of wood in the freezing waters of the North Atlantic, she stares straight up and we are treated to her view of this Hollywood sky—one where the stars on the right half of the scene trace the mirror image of the stars in the left half. How lazy can you get? To get it right would not have required a major realignment of the film's budget.

What's odd is that nobody would have known whether Cameron captured his plate and silverware patterns accurately. Whereas for about fifty bucks you can buy any one of a dozen programs for your home computer that will display the real sky for any time of day, any day of the year, any year of the millennium, and for any spot on Earth.

On one occasion, however, Cameron exercised artistic license commendably. After the *Titanic* sank, you see countless people (dead and alive) floating in the water. Of course, on this moonless night in the middle of the ocean, you would barely see the hand in front of your face. Cameron had to add illumination so that the viewer could follow the rest of the story. The lighting was soft and sensible, without obvious shadows indicating an embarrassing (and nonexistent) source of light.

This story actually has a happy ending. As many people know, James Cameron is a modern-day explorer, who does, in fact, value the scientific enterprise. His undersea expedition to the *Titanic* was one of many he has launched, and he served for many years on NASA's high-level Advisory Council. During a recent occasion in New York City, when he was honored by *Wired* magazine for his adventurous spirit, I was invited to dinner with the editors and Cameron himself. What better occasion to tell him of his errant ways with the *Titanic* sky. So after I whined for ten minutes on the subject, he replied, "The film, worldwide, has grossed over a billion dollars. Imagine how much more money it would have made had I gotten the night sky correct!"

I have never before been so politely, yet thoroughly, silenced. I meekly returned to my appetizer, mildly embarrassed for having raised the issue. Two months later, a phone call comes to my planetarium office. It was a computer visualization expert from a postproduction unit for James Cameron. He said that for their reissue of the film *Titanic*, in a Special Collector's Edition, they would be restoring some scenes and he was told I may have an accurate night sky they might want to use for this edition. Sure enough, I generated the right image of the night sky for every possible direction that Kate Winslet and Leonardo DiCaprio could turn their heads while the ship sank.

THE ONLY TIME I ever bothered to compose a letter complaining about a cosmic mistake was after I saw the 1991 romantic comedy *L.A. Story*, written and produced by Steve Martin. In this film, Martin uses the Moon to track time by showing its phase progressing from crescent to full. A big deal is not made of this fact. The Moon just hangs there in the sky from night to night. I applaud Martin's effort to engage the universe in his plot line, but this Hollywood moon grew in the wrong direction. Viewed from any location north of Earth's equator (Los Angeles qualifies), the Moon's illuminated surface grows from right to left.

When the Moon is a thin crescent, you can find the Sun 20 or 30 degrees to its right. As the Moon orbits Earth, the angle between it and the Sun grows, allowing more and more of its visible surface to be lit, reaching 100 percent frontal illumination at 180 degrees. (This monthly Earth-Sun-Moon configuration is known as syzygy, which reliably gives you a full Moon and, occasionally, a lunar eclipse.)

Steve Martin's moon grew from left to right. It grew backward.

My letter to Mr. Martin was polite and respectful, written on the assumption that he would want to know the cosmic truth. Alas, I received no reply, but then again, I was only in graduate school at the time, without a weighty letterhead to grab his attention.

Even the 1983 macho test-pilot epic *The Right Stuff* had plenty of the wrong stuff. In my favorite transgression, Chuck Yeager, the first to fly faster than the speed of sound, is shown ascending to 80,000 feet, setting yet another altitude and speed record. Ignoring the fact that the scene takes place in California's Mojave Desert, where clouds of any species are rare, as Yeager darts through the air you see puffy, white, alto-cumulus clouds whizzing by. This error would surely irk meteorologists because, in Earth's real atmosphere, these clouds would not be caught dead above 20,000 feet.

Without those visual props, I suppose the viewer would have no visceral idea of how fast the plane was moving. So I understand the motive. But the film's director, Philip Kaufman, was not without choices: Other kinds of clouds, such as cirrus, and the especially beautiful noctilucent clouds, do exist at very high altitudes. At some point in your life you have to learn that they exist.

The 1997 film *Contact*, inspired by Carl Sagan's 1983 novel of the same name, contains an especially embarrassing astro-gaffe. (I saw the movie and never read the book. But everyone who has read the book says, of course, that it's better than the movie.) *Contact* explores what might happen when humans find intelligent life in the galaxy and then establish contact with it. The heroine astrophysicist and alien hunter is actress Jodie Foster, who recites a fundamental line that contains mathematically impossible information. Just as she establishes her love interest in ex-priest Matthew McConaughey, seated with the largest radio telescope in the world behind them, she says to him with passion: "If there are 400 billion stars in the galaxy, and just one in a million had planets, and just one in a million of those had life, and just one in a million of those had intelligent life, that still leaves millions of planets to explore." Wrong. According to her numbers, that leaves 0.0000004 planets with intelligent life on them, which is a figure somewhat lower than "millions." No doubt that "one in a million" sounds better on screen than "one in ten," but you can't fake math.

Ms. Foster's recitation was not a gratuitous expression of math, it was an explicit recognition of the famous Drake equation, named for Astronomer Frank Drake who first calculated the likelihood of finding intelligent life in the galaxy based on a sequence of factors, starting with the total number of stars in the galaxy. For this reason, it was one of the most important scenes in the film. Who do we blame for the flub? Not the screenwriters, even though their words were spoken verbatim. I blame Jodie. As the lead actress, she forms the last line of defense against errors that creep into the lines she delivers. So she must bear some responsibility. Not only that, last I checked, she was a graduate of Yale University. Surely they teach arithmetic there.

During the 1970s and 1980s, the popular television soap opera *As The World Turns* portrayed sunrise during the opening credits and sunset during the closing credits, which, given the show's title, was a suitable cinematic gesture. Unfortunately, their sunrise was a sunset filmed in reverse. Nobody took the time to notice that for every day of the year in the Northern Hemisphere the Sun moves at an angle up and to the right of the spot on the horizon where it rises. At the end of the day, it descends across the sky at an angle down and to the right. The soap-opera sunrise showed the Sun moving toward the left as it rose. They obviously had gotten a piece of film showing a sunset and played it in reverse for the show's beginning. The producers were either too sleepy to wake up early and film the sunrise, or the sunrise was filmed in the Southern Hemisphere—after which the camera crew ran to the Northern Hemisphere to film sunset. Had they called their local astrophysicists, any one of us might have recommended that if they needed to save money, they could have shown the sunset in a mirror before they showed it running backward. This would have taken care of everybody's needs.

Of course, inexcusable astro-illiteracy extends beyond television, film, and paintings at the Louvre. The famous star-studded ceiling of New York City's Grand Central Terminal rises high above the countless busy commuters. I would be gripeless if the original designers had no pretense of portraying an authentic sky. But this three-acre canvas contains among its several hundred stars a dozen real constellations, each traced in their classical splendor, with the Milky Way flowing by, just where you're supposed to find it. Holding aside the sky's greenish color, which greatly resembles that of Sears household appliances from the 1950s, the sky is backward. Yes, backward. Turns out, this was common practice during the Renaissance, back when globe makers made celestial spheres. But in those cases, you, the viewer, stood in a mythical place "outside" of the sky, looking down, with Earth imagined to occupy the globe's center. This argument works well for spheres smaller than you, but fails miserably for 130-foot ceilings. And amid the backwardness, for reasons I have yet to divine, the stars of the constellation Orion are positioned forward, with Betelgeuse and Rigel correctly oriented.

Astrophysics is surely not the only science trod upon by underinformed artists. Naturalists have probably logged more gripes than we have. I can hear them now: "That's the wrong whale song for the species of whale they showed in the film." "Those plants are not native to that region." "Those rock formations have no relation to that terrain." "The sounds made by those geese are from a species that flies nowhere near that location." "They would have us believe it's the middle of the winter yet that maple tree still has all its leaves."

In my next life, what I plan to do is open a school for artistic science, where creative people can be accredited in their knowledge of the natural world. Upon graduating, they would be allowed to distort nature only in informed ways that advance their artistic needs. As the credits roll by, the director, producer, set designer, cinematographer, and whoever else was accredited would proudly list their membership with SCIPAL, the Society for Credible Infusion of Poetic and Artistic License.

SCIENCE

AND

GOD

WHEN WAYS OF KNOWING COLLIDE

IN THE BEGINNING

Physics describes the behavior of matter, energy, space, and time, and the interplay among them in the universe. From what scientists have been able to determine, all biological and chemical phenomena are ruled by what those four characters in our cosmic drama do to one another. And so everything fundamental and familiar to us Earthlings begins with the laws of physics.

In almost any area of scientific inquiry, but especially in physics, the frontier of discovery lives at the extremes of measurement. At the extremes of matter, such as the neighborhood of a black hole, you find gravity badly warping the surrounding space-time continuum. At the extremes of energy, you sustain thermonuclear fusion in the ten-million-degree cores of stars. And at every extreme imaginable, you get the outrageously hot, outrageously dense conditions that prevailed during the first few moments of the universe.

Daily life, we're happy to report, is wholly devoid of extreme physics. On a normal morning, you get out of bed, wander around the house, eat something, dash out the front door. And, by day's end, your loved ones fully expect you to look no different from the way you did when you left and to return home in one piece. But imagine arriving at the office, walking into an overheated conference room for an important 10:00 A.M. meeting, and suddenly los-

This essay was the winner of the 2005 Science Writing Award from the American Institute of Physics.

ing all your electrons—or worse yet, having every atom of your body fly apart. Or suppose you're sitting in your office trying to get some work done by the light of your desk lamp and somebody flicks on the overhead light, causing your body to bounce randomly from wall to wall until you're jack-in-the-boxed out the window. Or what if you went to a sumo wrestling match after work and saw the two spherical gentlemen collide, disappear, then spontaneously become two beams of light?

If those scenes played out daily, then modern physics wouldn't look so bizarre, knowledge of its foundations would flow naturally from our life experience, and our loved ones probably would never let us go to work. Back in the early minutes of the universe, though, that stuff happened all the time. To envision it, and understand it, one has no choice but to establish a new form of common sense, an altered intuition about how physical laws apply to extremes of temperature, density, and pressure.

Enter the world of $E = mc^2$.

Albert Einstein first published a version of this famous equation in 1905 in a seminal research paper titled "On the Electrodynamics of Moving Bodies." Better known as the special theory of relativity, the concepts advanced in that paper forever changed our notions of space and time. Einstein, then just 26 years old, offered further details about his tidy equation in a separate, remarkably short paper published later the same year: "Does the Inertia of a Body Depend on Its Energy Content?" To save you the effort of digging up the original article, designing an experiment, and testing the theory, the answer is "Yes." As Einstein wrote:

> If a body gives off the energy E in the form of radiation, its mass diminishes by E/c^2. . . . The mass of a body is a measure of its energy-content; if the energy changes by E, the mass changes in the same sense. (1952, p. 71)

Uncertain as to the truth of his statement, he then suggested:

It is not impossible that with bodies whose energy-content is variable to a high degree (e.g. with radium salts) the theory may be successfully put to the test. (1952, p. 71)

There you have it—the algebraic recipe for all occasions when you want to convert matter into energy or energy into matter. In those simple sentences, Einstein unwittingly gave astrophysicists a computational tool, $E = mc^2$, that extends their reach from the universe as it now is, all the way back to infinitesimal fractions of a second after its birth.

The most familiar form of energy is the photon, a massless, irreducible particle of light. You are forever bathed in photons: from the Sun, the Moon, and the stars to your stove, your chandelier, and your night-light. So why don't you experience $E = mc^2$ every day? The energy of visible-light photons falls far below that of the least massive subatomic particles. There is nothing else those photons can become, and so they live happy, relatively uneventful lives.

Want to see some action? Start hanging around gamma-ray photons that have some real energy—at least 200,000 times more than that of visible photons. You'll quickly get sick and die of cancer, but before that happens you'll see pairs of electrons—one matter, the other antimatter; one of many dynamic duos in the particle universe—pop into existence where photons once roamed. As you watch, you will also see matter-antimatter pairs of electrons collide, annihilating each other and creating gamma-ray photons once again. Increase the light's energy by a factor of another 2,000, and you now have gamma rays with enough energy to turn susceptible people into the Hulk. But pairs of these photons now have enough energy to spontaneously create the more massive neutrons, protons, and their antimatter partners.

High-energy photons don't hang out just anywhere. But the place needn't be imaginary. For gamma rays, almost any environment hotter than a few billion degrees will do just fine.

The cosmological significance of particles and energy packets transmuting into each other is staggering. Currently the temperature of our expanding universe, calculated from measurements of the microwave bath of light that pervades all of space, is a mere 2.73 degrees Kelvin. Like the photons of visible light, microwave photons are too cool to have any realistic ambitions to become a particle via $E = mc^2$; in fact, there are no known particles they can spontaneously become. Yesterday, however, the universe was a little bit smaller and a little bit hotter. The day before, it was smaller and hotter still. Roll the clocks backward some more—say, 13.7 billion years—and you land squarely in the primordial soup of the big bang, a time when the temperature of the cosmos was high enough to be astrophysically interesting.

The way space, time, matter, and energy behaved as the universe expanded and cooled from the beginning is one of the greatest stories ever told. But to explain what went on in that cosmic crucible, you must find a way to merge the four forces of nature into one, and find a way to reconcile two incompatible branches of physics: quantum mechanics (the science of the small) and general relativity (the science of the large).

Spurred by the successful marriage of quantum mechanics and electromagnetism in the mid-twentieth century, physicists set off on a race to blend quantum mechanics and general relativity (into a theory of quantum gravity). Although we haven't yet reached the finish line, we know exactly where the high hurdles are: during the "Planck era." That's the phase up to 10^{-43} seconds (one ten-million-trillion-trillion-trillionth of a second) after the beginning, and before the universe grew to 10^{-35} meters (one hundred-billion-trillion-trillionth of a meter) across. The German physicist Max Planck, after whom these unimaginably small quantities are named, introduced the idea of quantized energy in 1900 and is generally credited with being the father of quantum mechanics.

Not to worry, though. The clash between gravity and quantum mechanics poses no practical problem for the contemporary universe. Astrophysicists apply the tenets and tools of general relativity and quantum mechanics to very different classes of problems. But in the beginning, during the Planck era, the large was small, and there must have been a kind of shotgun wedding between the two. Alas, the vows exchanged during that ceremony continue to elude us, and so no (known) laws of physics describe with any confidence the behavior of the universe during the brief interregnum.

At the end of the Planck era, however, gravity wriggled loose from the other, still-unified forces of nature, achieving an independent identity nicely described by our current theories. As the universe aged through 10^{-35} seconds it continued to expand and cool, and what remained of the unified forces split into the electroweak and the strong nuclear forces. Later still, the electroweak force split into the electromagnetic and the weak nuclear forces, laying bare the four distinct forces we have come to know and love—with the weak force controlling radioactive decay, the strong force binding the nucleus, the electromagnetic force binding molecules, and gravity binding bulk matter. By now, the universe was a mere trillionth of a second old. Yet its transmogrified forces and other critical episodes had already imbued our universe with fundamental properties each worthy of its own book.

While the universe dragged on for its first trillionth of a second, the interplay of matter and energy was incessant. Shortly before, during, and after the strong and electroweak forces parted company, the universe was a seething ocean of quarks, leptons, and their antimatter siblings, along with bosons, the particles that enable their interactions. None of these particle families is thought to be divisible into anything smaller or more basic. Fundamental though they are, each comes in several species. The ordinary visible-light photon is a member of the boson family. The leptons most familiar to the nonphysicist are the electron and perhaps the neutrino; and the most familiar quarks are . . . well, there are no familiar quarks. Each species has been assigned an abstract name that serves no real philological, philosophical, or pedagogical purpose except to distinguish it from the others: up and down, strange and charmed, and top and bottom.

Bosons, by the way, are simply named after the Indian scientist

Satyendranath Bose. The word "lepton" derives from the Greek *leptos*, meaning "light" or "small." "Quark," however, has a literary and far more imaginative origin. The physicist Murray Gell-Mann, who in 1964 proposed the existence of quarks, and who at the time thought the quark family had only three members, drew the name from a characteristically elusive line in James Joyce's *Finnegans Wake*: "Three quarks for Muster Mark!" One thing quarks do have going for them: all their names are simple—something chemists, biologists, and geologists seem incapable of achieving when naming their own stuff.

Quarks are quirky beasts. Unlike protons, each with an electric charge of +1, and electrons, with a charge of −1, quarks have fractional charges that come in thirds. And you'll never catch a quark all by itself; it will always be clutching onto other quarks nearby. In fact, the force that keeps two (or more) of them together actually grows stronger the more you separate them—as if they were attached by some sort of subnuclear rubber band. Separate the quarks enough, the rubber band snaps and the stored energy summons $E = mc^2$ to create a new quark at each end, leaving you back where you started.

But during the quark-lepton era the universe was dense enough for the average separation between unattached quarks to rival the separation between attached quarks. Under those conditions, allegiance between adjacent quarks could not be unambiguously established, and they moved freely among themselves, in spite of being collectively bound to each other. The discovery of this state of matter, a kind of quark soup, was reported for the first time in 2002 by a team of physicists at the Brookhaven National Laboratories.

Strong theoretical evidence suggests that an episode in the very early universe, perhaps during one of the force splits, endowed the universe with a remarkable asymmetry, in which particles of matter barely outnumbered particles of antimatter by a billion-and-one to a billion. That small difference in population hardly got noticed amid the continuous creation, annihilation, and re-creation of quarks and antiquarks, electrons and antielectrons (better known as positrons), and neutrinos and antineutrinos. The odd man out had plenty of opportunities to find someone to annihilate with, and so did everybody else.

But not for much longer. As the cosmos continued to expand and cool, it became the size of the solar system, with a temperature dropping rapidly past a trillion degrees Kelvin.

A millionth of a second had passed since the beginning.

This tepid universe was no longer hot enough or dense enough to cook quarks, and so they all grabbed dance partners, creating a permanent new family of heavy particles called hadrons (from the Greek *hadros,* meaning "thick"). That quark-to-hadron transition soon resulted in the emergence of protons and neutrons as well as other, less familiar heavy particles, all composed of various combinations of quark species. The slight matter-antimatter asymmetry afflicting the quark-lepton soup now passed to the hadrons, but with extraordinary consequences.

As the universe cooled, the amount of energy available for the spontaneous creation of basic particles dropped. During the hadron era, ambient photons could no longer invoke $E = mc^2$ to manufacture quark-antiquark pairs. Not only that, the photons that emerged from all the remaining annihilations lost energy to the ever-expanding universe and dropped below the threshold required to create hadron-antihadron pairs. For every billion annihilations—leaving a billion photons in their wake—a single hadron survived. Those loners would ultimately get to have all the fun: serving as the source of galaxies, stars, planets, and people.

Without the billion-and-one to a billion imbalance between matter and antimatter, all mass in the universe would have annihilated, leaving a cosmos made of photons *and nothing else*—the ultimate let-there-be-light scenario.

By now, one second of time has passed.

The universe has grown to a few light-years across, about the distance from the Sun to its closest neighboring stars. At a billion degrees, it's still plenty hot—and still able to cook electrons, which, along with their positron counterparts, continue to pop in and out of existence. But in the ever-expanding, ever-cooling universe, their days (seconds, really) are numbered. What was true for hadrons is true for electrons: eventually only one electron in a billion survives. The rest get annihilated, together with their antimatter sidekicks the positrons, in a sea of photons.

Right about now, one electron for every proton has been "frozen" into existence. As the cosmos continues to cool—dropping below 100 million degrees—protons fuse with protons as well as with neutrons, forming atomic nuclei and hatching a universe in which 90 percent of these nuclei are hydrogen and 10 percent are helium, along with trace amounts of deuterium, tritium, and lithium.

Two minutes have now passed since the beginning.

Not for another 380,000 years does much happen to our particle soup. Throughout these millennia the temperature remains hot enough for electrons to roam free among the photons, batting them to and fro.

But all this freedom comes to an abrupt end when the temperature of the universe falls below 3,000 degrees Kelvin (about half the temperature of the Sun's surface), and all the electrons combine with free nuclei. The marriage leaves behind a ubiquitous bath of visible-light photons, completing the formation of particles and atoms in the primordial universe.

As the universe continues to expand, its photons continue to lose energy, dropping from visible light to infrared to microwaves.

As we will soon discuss in more detail, everywhere astrophysicists look we find an indelible fingerprint of 2.73-degree microwave photons, whose pattern on the sky retains a memory of the distribution of matter just before atoms formed. From this we can deduce many things, including the age and shape of the universe. And although atoms are now part of daily life, Einstein's equilibrious equation still has plenty of work to do—in particle accelerators, where matter-antimatter particle pairs are created routinely from energy fields; in the core of the Sun, where 4.4 million tons of matter are converted into energy every second; and in the cores of every other star.

It also manages to occupy itself near black holes, just outside their event horizons, where particle-antiparticle pairs can pop into existence at the expense of the black hole's formidable gravitational energy. Stephen Hawking first described that process in 1975, showing that the mass of a black hole can slowly evaporate by this mechanism. In other words, black holes are not entirely black. Today the phenomenon is known as Hawking radiation and is a reminder of the continued fertility of $E = mc^2$.

But what happened before all this? What happened before the beginning?

Astrophysicists have no idea. Or, rather, our most creative ideas have little or no grounding in experimental science. Yet certain types of religious people tend to assert, with a tinge of smugness, that *something* must have started it all: a force greater than all others, a source from which everything issues. A prime mover.

In the mind of such a person, that something is, of course, God.

But what if the universe was always there, in a state or condition we have yet to identify—a multiverse, for instance? Or what if the universe, like its particles, just popped into existence from nothing?

Such replies usually satisfy nobody. Nonetheless, they remind us that ignorance is the natural state of mind for a research scientist on the ever-shifting frontier. People who believe they are ignorant of nothing have neither looked for, nor stumbled upon, the boundary between what is known and unknown in the cosmos. And therein lies a fascinating dichotomy. "The universe always was" goes unrecognized as a legitimate answer to "What was around before the beginning?" But for many religious people, the answer "God always was" is the obvious and pleasing answer to "What was around before God?"

No matter who you are, engaging in the quest to discover where and how things began tends to induce emotional fervor—as if knowing the beginning bestows upon you some form of fellowship with, or perhaps governance over, all that comes later. So what is true for life itself is no less true for the universe: knowing where you came from is no less important than knowing where you are going.

HOLY WARS

At nearly every public lecture that I give on the universe, I try to reserve adequate time at the end for questions. The succession of subjects is predictable. First, the questions relate directly to the lecture. They next migrate to sexy astrophysical subjects such as black holes, quasars, and the big bang. If I have enough time left over to answer all questions, and if the talk is in America, the subject eventually reaches God. Typical questions include, "Do scientists believe in God?" "Do you believe in God?" "Do your studies in astrophysics make you more or less religious?"

Publishers have come to learn that there is a lot of money in God, especially when the author is a scientist and when the book title includes a direct juxtaposition of scientific and religious themes. Successful books include Robert Jastrow's *God and the Astronomers*, Leon M. Lederman's *The God Particle*, Frank J. Tipler's *The Physics of Immortality: Modern Cosmology, God, and the Resurrection of the Dead*, and Paul Davies's two works *God and the New Physics* and *The Mind of God*. Each author is either an accomplished physicist or astrophysicist and, while the books are not strictly religious, they encourage the reader to bring God into conversations about astrophysics. Even the late Stephen Jay Gould, a Darwinian pitbull and devout agnostic, joined the title parade with his work *Rock of Ages: Science and Religion in the Fullness of Life*. The financial success of these published works indicates that you get bonus dollars from the American public if you are a scientist who openly talks about God.

After the publication of *The Physics of Immortality*, which suggested whether the law of physics could allow you and your soul to exist long after you are gone from this world, Tipler's book tour included many well-paid lectures to Protestant religious groups. This lucrative subindustry has further blossomed in recent years due to efforts made by the wealthy founder of the Templeton investment fund, Sir John Templeton, to find harmony and consilience between science and religion. In addition to sponsoring workshops and conferences on the subject, the Templeton Foundation seeks out widely published religion-friendly scientists to receive an annual award whose cash value exceeds that of the Nobel Prize.

Let there be no doubt that as they are currently practiced, there is no common ground between science and religion. As was thoroughly documented in the nineteenth-century tome *A History of the Warfare of Science with Theology in Christendom*, by the historian and onetime president of Cornell University Andrew D. White, history reveals a long and combative relationship between religion and science, depending on who was in control of society at the time. The claims of science rely on experimental verification, while the claims of religions rely on faith. These are irreconcilable approaches to knowing, which ensures an eternity of debate wherever and whenever the two camps meet. Although just as in hostage negotiations, it's probably best to keep both sides talking to each other.

The schism did not come about for want of earlier attempts to bring the two sides together. Great scientific minds, from Claudius Ptolemy of the second century to Isaac Newton of the seventeenth, invested their formidable intellects in attempts to deduce the nature of the universe from the statements and philosophies contained in religious writings. Indeed, by the time of his death, Newton had penned more words about God and religion than about the laws of physics, which included futile attempts to invoke the biblical chronology to understand and predict events in the natural world. Had any of these efforts succeeded, science and religion today might be largely indistinguishable.

The argument is simple. I have yet to see a successful prediction about the physical world that was inferred or extrapolated from the content of any religious document. Indeed, I can make an even stronger statement. Whenever people have tried to make accurate predictions about the physical world using religious documents they have been famously wrong. By a prediction, I mean a precise statement about the untested behavior of objects or phenomena in the natural world, logged *before* the event takes place. When your model predicts something only after it has happened then you have instead made a "postdiction." Postdictions are the backbone of most creation myths and, of course, of the *Just So Stories* of Rudyard Kipling, where explanations of everyday phenomena explain what is already known. In the business of science, however, a hundred postdictions are barely worth a single successful prediction.

TOPPING THE LIST of religious predictions are the perennial claims about when the world will end, none of which have yet proved true. A harmless enough exercise. But other claims and predictions have actually stalled or reversed the progress of science. We find a leading example in the trial of Galileo (which gets my vote for the trial of the millennium) where he showed the universe to be fundamentally different from the dominant views of the Catholic Church. In all fairness to the Inquisition, however, an Earth-centered universe made lots of sense observationally. With a full complement of epicycles to explain the peculiar motions of the planets against the background stars, the time-honored, Earth-centered model had conflicted with no known observations. This remained true long after Copernicus introduced his Sun-centered model of the universe a century earlier. The Earth-centric model was also aligned with the teachings of the Catholic Church and prevailing interpretations of the Bible, wherein Earth is unambiguously created before the Sun and the Moon as described in the first several verses of Genesis. If you were created first, then you must be in the center of all motion. Where else could you be? Furthermore, the Sun and Moon themselves were also presumed to be smooth orbs. Why would a perfect, omniscient deity create anything else?

All this changed, of course, with the invention of the telescope and Galileo's observations of the heavens. The new optical device revealed aspects of the cosmos that strongly conflicted with people's conceptions of an Earth-centered, blemish-free, divine universe: The Moon's surface was bumpy and rocky; the Sun's surface had spots that moved across its surface; Jupiter had moons of its own that orbited Jupiter and not Earth; and Venus went through phases, just like the Moon. For his radical discoveries, which shook Christendom—and for being a pompous jerk about it—Galileo was put on trial, found guilty of heresy, and sentenced to house arrest. This was mild punishment when one considers what happened to the monk Giordano Bruno. A few decades earlier Bruno had been found guilty of heresy, and then burned at the stake, for suggesting that Earth may not be the only place in the universe that harbors life.

I do not mean to imply that competent scientists, soundly following the scientific method, have not also been famously wrong. They have. Most scientific claims made on the frontier will ultimately be disproved, due primarily to bad or incomplete data, and occasionally to blunder. But the scientific method, which allows for expeditions down intellectual dead ends, also promotes ideas, models, and predictive theories that can be spectacularly correct. No other enterprise in the history of human thought has been as successful at decoding the ways and means of the universe.

Science is occasionally accused of being a closed-minded or stubborn enterprise. Often people make such accusations when they see scientists swiftly discount astrology, the paranormal, Sasquatch sightings, and other areas of human interest that routinely fail double-blind tests or that possess a dearth of reliable evidence. But don't be offended. Scientists apply this same level of skepticism to ordinary claims in the professional research journals. The standards are identical. Look what happened when the Utah chemists B. Stanley Pons and Martin Fleischmann claimed in a press conference to have created "cold" nuclear fusion on their laboratory table. Scientists acted swiftly and skeptically. Within days of the announcement it was clear that no one could replicate the cold fusion results that Pons and Fleischmann claimed. Their work was summarily dismissed. Similar plot lines unfold almost daily (minus the press conferences) for nearly every new scientific claim. The ones you hear about tend to be only those that could affect the economy.

WITH SCIENTISTS EXHIBITING such strong levels of skepticism, some people may be surprised to learn that scientists heap their largest rewards and praises upon those who do, in fact, discover flaws in established paradigms. These same rewards also go to those who create new ways to understand the universe. Nearly all famous scientists, pick your favorite one, have been so praised in their own lifetimes. This path to success in one's professional career is antithetical to almost every other human establishment—especially to religion.

None of this is to say that the world does not contain religious scientists. In a recent survey of religious beliefs among math and science professionals (Larson and Witham 1998), 65 percent of the mathematicians (the highest rate) declared themselves to be religious, as did 22 percent of the physicists and astronomers (the lowest rate). The national average among all scientists was around 40 percent and has remained largely unchanged over the past century. For reference, about 90 percent of the American public claims to be religious (among the highest in Western society), so either nonreligious people are drawn to science or studying science makes you less religious.

But what of those scientists who are religious? Successful researchers do not get their science from their religious beliefs. On the other hand, the methods of science currently have little or nothing to contribute to ethics, inspiration, morals, beauty, love, hate, or aesthetics. These are vital elements of civilized life and are central to the concerns of nearly every religion. What it all means is that for many scientists there is no conflict of interest.

As we will soon see in detail, when scientists do talk about God, they typically invoke him at the boundaries of knowledge where we should be most humble and where our sense of wonder is greatest.

Can one grow tired of wonderment?

In the thirteenth century, Alfonso the Wise (Alfonso X), the king of Spain, who also happened to be an accomplished academician, was frustrated by the complexity of Ptolemy's epicycles accounting for the geocentric universe. Being less humble than others on the frontier, Alfonso once mused, "Had I been around at the creation, I would have given some useful hints for the better ordering of the universe" (Carlyle 2004, Book II, Chapter VII).

In full agreement with King Alfonso's frustrations with the universe, Albert Einstein noted in a letter to a colleague, "If God created the world, his primary worry was certainly not to make its understanding easy for us" (1954). When Einstein could not figure out how or why a deterministic universe could require the probabilistic formalisms of quantum mechanics, he mused, "It is hard to sneak a look at God's cards. But that He would choose to play dice with the world . . . is something that I cannot believe for a single moment" (Frank 2002, p. 208). When an experimental result was shown to Einstein that, if correct, would have disproved his new theory of gravity Einstein commented, "The Lord is subtle, but malicious He is not" (Frank 2002, p. 285). The Danish physicist Niels Bohr, a contemporary of Einstein, heard one too many of Einstein's God-remarks and declared that Einstein should stop telling God what to do! (Gleick 1999)

Today, you hear the occasional astrophysicist (maybe one in a hundred) publicly invoke God when asked where did all our laws of physics come from or what was around before the big bang. As we have come to anticipate, these questions comprise the modern frontier of cosmic discovery and, at the moment, they transcend the answers our available data and theories can supply. Some promising ideas, such as inflationary cosmology and string theory, already exist. These could ultimately provide the answers to those questions, further pushing back our boundary of awe.

My personal views are entirely pragmatic and partly resonate with those of Galileo who, during his trial, is credited with saying, "The Bible tells you how to go to heaven, not how the heavens go" (Drake 1957, p. 186). Galileo further noted, in a 1615 letter to the Grand Duchess of Tuscany, "In my mind God wrote two books. The first book is the Bible, where humans can find the answers to their questions on values and morals. The second book of God is the book of nature, which allows humans to use observation and experiment to answer our own questions about the universe" (Drake 1957, p. 173).

I simply go with what works. And what works is the healthy skepticism embodied in scientific method. Believe me, if the Bible had ever been shown to be a rich source of scientific answers and understanding, we would be mining it daily for cosmic discovery. Yet my vocabulary of scientific inspiration strongly overlaps with that of religious enthusiasts. I, like others, am humbled in the presence of the objects and phenomena of our universe. And I go misty with admiration for its splendor. But I do so knowing and accepting that if I propose a God who graces our valley of unknowns, the day may come, empowered by the advance of science, when no more valleys remain.

THE PERIMETER OF IGNORANCE

Writing in centuries past, many scientists felt compelled to wax poetic about cosmic mysteries and God's handiwork. Perhaps one should not be surprised at this: most scientists back then, as well as many scientists today, identify themselves as spiritually devout.

But a careful reading of older texts, particularly those concerned with the universe itself, shows that the authors invoke divinity only when they reach the boundaries of their understanding. They appeal to a higher power only when staring into the ocean of their own ignorance. They call on God only from the lonely and precarious edge of incomprehension. Where they feel certain about their explanations, however, God gets hardly a mention.

Let's start at the top. Isaac Newton was one of the greatest intellects the world has ever seen. His laws of motion and his universal law of gravitation, conceived in the mid-seventeenth century, account for cosmic phenomena that had eluded philosophers for millennia. Through those laws, one could understand the gravitational attraction of bodies in a system, and thus come to understand orbits.

Newton's law of gravity enables you to calculate the force of attraction between any two objects. If you introduce a third object, then each one attracts the other two, and the orbits they trace become much harder to compute. Add another object, and another, and another, and soon you have the planets in our solar system. Earth and the Sun pull on each other, but Jupiter also pulls on

Earth, Saturn pulls on Earth, Mars pulls on Earth, Jupiter pulls on Saturn, Saturn pulls on Mars, and on and on.

Newton feared that all this pulling would render the orbits in the solar system unstable. His equations indicated that the planets should long ago have either fallen into the Sun or flown the coop—leaving the Sun, in either case, devoid of planets. Yet the solar system, as well as the larger cosmos, appeared to be the very model of order and durability. So Newton, in his greatest work, the *Principia*, concludes that God must occasionally step in and make things right:

> *The six primary Planets are revolv'd about the Sun, in circles concentric with the Sun, and with motions directed towards the same parts, and almost in the same plane. . . . But it is not to be conceived that mere mechanical causes could give birth to so many regular motions. . . . This most beautiful System of the Sun, Planets, and Comets, could only proceed from the counsel and dominion of an intelligent and powerful Being.* (1992, p. 544)

In the *Principia*, Newton distinguishes between hypotheses and experimental philosophy, and declares, "Hypotheses, whether metaphysical or physical, whether of occult qualities or mechanical, have no place in experimental philosophy" (p. 547). What he wants is data, "inferr'd from the phænomena." But in the absence of data, at the border between what he could explain and what he could only honor—the causes he could identify and those he could not—Newton rapturously invokes God:

> *Eternal and Infinite, Omnipotent and Omniscient; . . . he governs all things, and knows all things that are or can be done. . . . We know him only by his most wise and excellent contrivances of things, and final causes; we admire him for his perfections; but we reverence and adore him on account of his dominion.* (p. 545)

A century later, the French astronomer and mathematician Pierre-Simon Laplace confronted Newton's dilemma of unstable orbits head-on. Rather than view the mysterious stability of the solar system as the unknowable work of God, Laplace declared it a scientific challenge. In his multipart masterpiece, *Traité de mécanique céleste*, the first volume of which appeared in 1799, Laplace demonstrates that the solar system is stable over periods of time longer than Newton could predict. To do so, Laplace pioneered a new kind of mathematics called perturbation theory, which enabled him to examine the cumulative effects of many small forces. According to an oft-repeated but probably embellished account, when Laplace gave a copy of *Traité de mécanique céleste* to his physics-literate friend Napoleon Bonaparte, Napoleon asked him what role God played in the construction and regulation of the heavens. "Sire," Laplace replied, "I had no need of that hypothesis" (DeMorgan 1872).

LAPLACE NOTWITHSTANDING, plenty of scientists besides Newton have called on God—or the gods—wherever their comprehension fades to ignorance. Consider the second-century A.D. Alexandrian astronomer Ptolemy. Armed with a description, but no real understanding, of what the planets were doing up there, he could not contain his religious fervor and scribbled this note in the margin of his *Almagest*:

> I know that I am mortal by nature, and ephemeral; but when I trace, at my pleasure, the windings to and fro of the heavenly bodies, I no longer touch Earth with my feet: I stand in the presence of Zeus himself and take my fill of ambrosia.

Or consider the seventeenth-century Dutch astronomer Christiaan Huygens, whose achievements include constructing the first working pendulum clock and discovering the rings of Saturn. In his charming book *The Celestial Worlds Discover'd*, posthumously published in 1698, most of the opening chapter celebrates all that was then known of planetary orbits, shapes, and sizes, as well as the planets' relative brightness and presumed rockiness. The book even includes foldout charts illustrating the structure of the solar system. God is absent from this discussion—even though a mere century earlier, before Newton's achievements, planetary orbits were supreme mysteries.

Celestial Worlds also brims with speculations about life in the solar system, and that's where Huygens raises questions to which he has no answer. That's where he mentions the biological conundrums of the day, such as the origin of life's complexity. And sure enough, because seventeenth-century physics was more advanced than seventeenth-century biology, Huygens invokes the hand of God only when he talks about biology:

> *I suppose no body will deny but that there's somewhat more of Contrivance, somewhat more of Miracle in the production and growth of Plants and Animals than in lifeless heaps of inanimate Bodies. . . . For the finger of God, and the Wisdom of Divine Providence, is in them much more clearly manifested than in the other.* (p. 20)

Today secular philosophers call that kind of divine invocation "God of the gaps"—which comes in handy, because there has never been a shortage of gaps in people's knowledge.

AS REVERENT AS Newton, Huygens, and other great scientists of earlier centuries may have been, they were also empiricists. They did not retreat from the conclusions their evidence forced them to draw, and when their discoveries conflicted with prevailing articles of faith, they upheld the discoveries. That doesn't mean it was easy: sometimes they met fierce opposition, as did Galileo, who had to defend his telescopic evidence against formidable objections drawn from both scripture and "common" sense.

Galileo clearly distinguished the role of religion from the role of science. To him, religion was the service of God and the salvation of souls, whereas science was the source of exact observations and demonstrated truths. In his 1615 letter to the Grand Duchess Christina of Tuscany he leaves no doubt about where he stood on the literal word of the Holy Writ:

> *In expounding the Bible if one were always to confine oneself to the unadorned grammatical meaning, one might fall into error. . . .*
>
> *Nothing physical which . . . demonstrations prove to us, ought to be called in question (much less condemned) upon the testimony of biblical passages which may have some different meaning beneath their words. . . .*
>
> *I do not feel obliged to believe that the same God who has endowed us with senses, reason and intellect has intended us to forgo their use.*
> (Venturi 1818, p. 222)

A rare exception among scientists, Galileo saw the unknown as a place to explore rather than as an eternal mystery controlled by the hand of God.

As long as the celestial sphere was generally regarded as the domain of the divine, the fact that mere mortals could not explain its workings could safely be cited as proof of the higher wisdom and power of God. But beginning in the sixteenth century, the work of Copernicus, Kepler, Galileo, and Newton—not to mention Maxwell, Heisenberg, Einstein, and everybody else who discovered fundamental laws of physics—provided rational explanations for an increasing range of phenomena. Little by little, the universe was subjected to the methods and tools of science, and became a demonstrably knowable place.

THEN, IN WHAT amounts to a stunning yet unheralded philosophical inversion, throngs of ecclesiastics and scholars began to declare that it was the laws of physics themselves that served as proof of the wisdom and power of God.

One popular theme of the seventeenth and eighteenth centuries was the "clockwork universe"—an ordered, rational, predictable mechanism fashioned and run by God and his physical laws. The early telescopes, which all relied on visible light, did little to undercut that image of an ordered system. The Moon revolved around Earth. Earth and other planets rotated on their axes and revolved around the Sun. The stars shone. The nebulae floated freely in space.

Not until the nineteenth century was it evident that visible light is just one band of a broad spectrum of electromagnetic radiation—the band that human beings just happen to see. Infrared was discovered in 1800, ultraviolet in 1801, radio waves in 1888, x-rays in 1895, and gamma rays in 1900. Decade by decade in the following century, new kinds of telescopes came into use, fitted with detectors that could "see" these formerly invisible parts of the electromagnetic spectrum. Now astrophysicists began to unmask the true character of the universe.

Turns out that some celestial bodies give off more light in the invisible bands of the spectrum than in the visible. And the invisible light picked up by the new telescopes showed that mayhem abounds in the cosmos: monstrous gamma-ray bursts, deadly pulsars, matter-crushing gravitational fields, matter-hungry black holes that flay their bloated stellar neighbors, newborn stars igniting within pockets of collapsing gas. And as our ordinary, optical telescopes got bigger and better, more mayhem emerged: galaxies that collide and cannibalize each other, explosions of supermassive stars, chaotic stellar and planetary orbits. And as noted earlier our own cosmic neighborhood—the inner solar system—turned out to be a shooting gallery, full of rogue asteroids and comets that collide with planets from time to time. Occasionally they've even wiped out stupendous masses of Earth's flora and fauna. The evidence all points to the fact that we occupy not a well-mannered clockwork universe, but a destructive, violent, and hostile zoo.

Of course, Earth can be bad for your health too. On land, grizzly bears want to maul you; in the oceans, sharks want to eat you. Snowdrifts can freeze you, deserts dehydrate you, earthquakes bury you, volcanoes incinerate you. Viruses can infect you, parasites suck your vital fluids, cancers take over your body, congenital diseases force an early death. And even if you have the good luck to be healthy, a swarm of locusts could devour your crops, a tsunami could wash away your family, or a hurricane could blow apart your town.

SO THE UNIVERSE wants to kill us all. But, as we have before, let's ignore that complication for the moment.

Many, perhaps countless, questions hover at the front lines of science. In some cases, answers have eluded the best minds of our species for decades or even centuries. And in contemporary America, the notion that a higher intelligence is the single answer to all enigmas has been enjoying a resurgence. This present-day version of God of the gaps goes by a fresh name: "intelligent design." The term suggests that some entity, endowed with a mental capacity far greater than the human mind can muster, created or enabled all the things in the physical world that we cannot explain through scientific methods.

An interesting hypothesis.

But why confine ourselves to things too wondrous or intricate for us to understand, whose existence and attributes we then credit to a superintelligence? Instead, why not tally all those things whose design is so clunky, goofy, impractical, or unworkable that they reflect the absence of intelligence?

Take the human form. We eat, drink, and breathe through the same hole in the head, and so, despite Henry J. Heimlich's eponymous maneuver, choking is the fourth leading cause of "unintentional injury death" in the United States. How about drowning, the fifth leading cause? Water covers almost three-quarters of Earth's surface, yet we are land creatures—submerge your head for just a few minutes and you die.

Or take our collection of useless body parts. What good is the pinky toenail? How about the appendix, which stops functioning after childhood and thereafter serves only as the source of appendicitis? Useful parts, too, can be problematic. I happen to like my knees, but nobody ever accused them of being well protected from bumps and bangs. These days, people with problem knees can get them surgically replaced. As for our pain-prone spine, it may be a while before someone finds a way to swap that out.

How about the silent killers? High blood pressure, colon cancer, and diabetes each cause tens of thousands of deaths in the U.S. every year, but it's possible not to know you're afflicted until your coroner tells you so. Wouldn't it be nice if we had built-in biogauges to warn us of such dangers well in advance? Even cheap cars, after all, have engine gauges.

And what comedian configured the region between our legs—an entertainment complex built around a sewage system?

The eye is often held up as a marvel of biological engineering. To the astrophysicist, though, it's only a so-so detector. A better one would be much more sensitive to dark things in the sky and to all the invisible parts of the spectrum. How much more breathtaking sunsets would be if we could see ultraviolet and infrared. How useful it would be if, at a glance, we could see every source of microwaves in the environment, or know which radio station transmitters were active. How helpful it would be if we could spot police radar detectors at night.

Think how easy it would be to navigate an unfamiliar city if we, like birds, could always tell which way was north because of the magnetite in our heads. Think how much better off we'd be if we had gills as well as lungs, how much more productive if we had six arms instead of two. And if we had eight, we could safely drive a car while simultaneously talking on a cell phone, changing the radio station, applying makeup, sipping a drink, and scratching our left ear.

Stupid design could fuel a movement unto itself. It may not be nature's default, but it's ubiquitous. Yet people seem to enjoy think-ing that our bodies, our minds, and even our universe represent pinnacles of form and reason. Maybe it's a good antidepressant to think so. But it's not science—not now, not in the past, not ever.

ANOTHER PRACTICE THAT isn't science is *embracing* ignorance. Yet it's fundamental to the philosophy of intelligent design: I don't know what this is. I don't know how it works. It's too complicated for me to figure out. It's too complicated for any human being to figure out. So it must be the product of a higher intelligence.

What do you do with that line of reasoning? Do you just cede the solving of problems to someone smarter than you, someone who's not even human? Do you tell students to pursue only questions with easy answers?

There may be a limit to what the human mind can figure out about our universe. But how presumptuous it would be for me to claim that if I can't solve a problem, neither can any other person who has ever lived or who will ever be born. Suppose Galileo and Laplace had felt that way? Better yet, what if Newton had *not*? He might then have solved Laplace's problem a century earlier, making it possible for Laplace to cross the next frontier of ignorance.

Science is a philosophy of discovery. Intelligent design is a phi-losophy of ignorance. You cannot build a program of discovery on the assumption that nobody is smart enough to figure out the answer to a problem. Once upon a time, people identified the god Neptune as the source of storms at sea. Today we call these storms hurricanes. We know when and where they start. We know what drives them. We know what mitigates their destructive power. And anyone who has studied global warming can tell you what makes them worse. The only people who still call hurricanes "acts of God" are the people who write insurance forms.

TO DENY OR ERASE the rich, colorful history of scientists and other thinkers who have invoked divinity in their work would be intellectually dishonest. Surely there's an appropriate place for intelligent design to live in the academic landscape. How about the history of religion? How about philosophy or psychology? The one place it doesn't belong is the science classroom.

If you're not swayed by academic arguments, consider the financial consequences. Allow intelligent design into science textbooks, lecture halls, and laboratories, and the cost to the frontier of scientific discovery—the frontier that drives the economies of the future—would be incalculable. I don't want students who could make the next major breakthrough in renewable energy sources or space travel to have been taught that anything they don't understand, and that nobody yet understands, is divinely constructed and therefore beyond their intellectual capacity. The day that happens, Americans will just sit in awe of what we don't understand, while we watch the rest of the world boldly go where no mortal has gone before.

REFERENCES

Modern publications of historic texts are listed when available.

Aristotle. 1943. *On Man in the Universe*. New York: Walter J. Black.

Aronson, A., and T. Ludlam, eds. 2005. *Hunting the Quark Gluon Plasma: Results from the First 3 Years at the Relativistic Heavy Ion Collider (RHIC)*. Upton, NY: Brookhaven National Laboratory. Formal Report: BNL -73847.

Atkinson, R. 1931. Atomic Synthesis and Stellar Energy. *Astrophysical Journal* 73: 250–95.

Aveni, Anthony. 1989. *Empires of Time*. New York: Basic Books.

Baldry, K., and K. Glazebrook. 2002. The 2dF Galaxy Redshift Survey: Constraints on Cosmic Star-Formation History from the Cosmic Spectrum. *Astrophysical Journal* 569: 582.

Barrow, John D. 1988. *The World within the World*. Oxford: Clarendon Press.

[Biblical passages] *The Holy Bible*. 1611. King James Translation.

Brewster, David. 1860. *Memoirs of the Life, Writings, and Discoveries of Sir Isaac Newton*, vol. 2. Edinburgh: Edmonston.

[Bruno, Giordano] Dorothea Waley Singer. 1950. *Giordano Bruno* (containing *On the Infinite Universe and Worlds [1584]*). New York: Henry Schuman.

Burbidge, E. M.; Geoffrey. R. Burbidge, William Fowler, and Fred Hoyle. 1957. The Synthesis of the Elements in Stars. *Reviews of Modern Physics* 29:15.

Carlyle, Thomas. 2004. *History of Frederick the Great* [1858]. Kila, MT: Kessinger Publishing.

[Central Bureau for Astronomical Telegrams] Brian Marsden, ed. 1998. Cambridge, MA: Center for Astrophysics, March 11, 1998.

Chaucer, Geoffrey. 1964. Prologue. *The Canterbury Tales* [1387]. New York: Modern Library.

Clarke, Arthur C. 1961. *A Fall of Moondust.* New York: Harcourt.

Clerke, Agnes M. 1890. *The System of the Stars.* London: Longmans, Green, & Co.

Comte, Auguste. 1842. *Coups de la Philosophie Positive,* vol. 2. Paris: Bailliere.

―――. 1853. *The Positive Philosophy of Auguste Compte,* London: J. Chapman.

Copernicus, Nicolaus. 1617. *De Revolutionibus Orbium Coelestium (Latin),* 3rd ed. Amsterdam: Wilhelmus Iansonius.

―――. 1999. *On the Revolutions of the Heavenly Sphere (English).* Norwalk, CT: Easton Press.

Darwin, Charles. 1959. Letter to J. D. Hooker, February 8, 1874. In *The Life and Letters of Charles Darwin.* New York: Basic Books.

―――. 2004. *The Origin of Species.* Edison, NJ: Castle Books.

DeMorgan, A. 1872. *Budget of Paradoxes.* London: Longmans Green & Co.

de Vaucouleurs, Gerard. 1983. Personal communication.

Doppler, Christian. 1843. On the Coloured Light of the Double Stars and Certain Other Stars of the Heavens. Paper delivered to the Royal Bohemian Society, May 25, 1842. *Abhandlungen der Königlich Böhmischen Gesellschaft der Wissenschaften,* Prague, 2: 465.

Eddington, Sir Arthur Stanley. 1920. *Nature* 106:14.

―――. 1926. *The Internal Constitution of the Stars.* Oxford, UK: Oxford Press.

Einstein, Albert. 1952. *The Principle of Relativity* [1923]. New York: Dover Publications.

―――. 1954. Letter to David Bohm. February 10. Einstein Archive 8-041.

[Einstein, Albert] James Gleick. 1999. Einstein, *Time,* December 31.

[Einstein, Albert] Phillipp Frank. 2002. *Einstein, His Life and Times* [1947]. Trans. George Rosen. New York: Da Capo Press.

Faraday, Michael. 1855. *Experimental Researches in Electricity.* London: Taylor.

Ferguson, James. 1757. *Astronomy Explained on Sir Isaac Newton's Principles,* 2nd ed. London: Globe.

Feynman, Richard. 1968. What Is Science. *The Physics Teacher* 7, no. 6: 313–20.

―――. 1994. *The Character of Physical Law.* New York: The Modern Library.

Forbes, George. 1909. *History of Astronomy.* London: Watts & Co.

Fraunhofer, Joseph von. 1898. *Prismatic and Diffraction Spectra.* Trans. J. S. Ames. New York: Harper & Brothers.

[Frost, Robert] Edward Connery Lathem, ed. 1969. *The Poetry of Robert Frost: The Collected Poems, Complete and Unabridged.* New York: Henry Holt and Co.

Galen. 1916. *On the Natural Faculties* [c. 180]. Trans. J. Brock. Cambridge, MA: Harvard University Press.

[Galileo, Galilei] Stillman Drake. 1957. *Discoveries and Opinions of Galileo.* New York: Doubleday Anchor Books.

Galileo, Galilei. 1744. *Opera.* Padova: Nella Stamperia.

———. 1954. *Dialogues Concerning Two New Sciences.* New York: Dover Publications.

———. 1989. *Sidereus Nucius* [1610]. Chicago: University of Chicago Press.

Gehrels, Tom, ed. 1994. *Hazards Due to Comets and Asteroids.* Tucson: University of Arizona Press.

Gillet, J. A., and W. J. Rolfe. 1882. *The Heavens Above.* New York: Potter Ainsworth & Co.

Gregory, Richard. 1923. *The Vault of Heaven.* London: Methuen & Co.

[Harrison, John] Dava Sobel. 2005. *Longitude.* New York: Walker & Co.

Hassan, Z., and Lui, eds. 1984. *Ideas and Realities: Selected Essays of Abdus Salaam.* Hackensack, NJ: World Scientific.

Heron of Alexandria. *Pneumatica* [c. 60].

Hertz, Heinrich. 1900. *Electric Waves.* London: Macmillan and Co.

Hubble Heritage Team. *Hubble Heritage Images.* http://heritage.stsci.edu.

Hubble, Edwin P. 1936. *Realm of the Nebulae.* New Haven, CT: Yale University Press.

———. 1954. *The Nature of Science.* San Marino, CA: Huntington Library.

Huygens, Christiaan. 1659. *Systema Saturnium (Latin).* Hagae-Comitis: Adriani Vlacq.

———. 1698. [*Cosmotheoros,*] *The Celestial Worlds Discover'd (English).* London: Timothy Childe.

Impey, Chris, and William K. Hartmann. 2000. *The Universe Revealed.* New York: Brooks Cole.

Johnson, David. 1991. *V-1, V-2: Hitler's Vengeance on London*. London: Scarborough House.

Kant, Immanuel. 1969. *Universal Natural History and Theory of the Heavens* [1755]. Ann Arbor: University of Michigan.

Kapteyn, J. C. 1909. On the Absorption of Light in Space. *Contrib. from the Mt. Wilson Solar Observatory,* no. 42, *Astrophysical Journal* (offprint), Chicago: University of Chicago Press.

Kelvin, Lord. 1901, Nineteenth Century Clouds over the Dynamical Theory of Heat and Light. In *London Philosophical Magazine and Journal of Science* 2, 6th Series, p. 1. Newcastle, UK: Literary and Philosophical Society.

———. 1904. *Baltimore Lectures*. Cambridge, UK: C. J. Clay and Sons.

Kepler, Johannes. 1992. *Astronomia Nova* [1609]. Trans. W. H. Donahue. Cambridge, UK: Cambridge University Press.

———. 1997. *The Harmonies of the World* [1619]. Trans. Juliet Field. Philadelphia: American Philosophical Society.

Lang, K. R., and O. Gingerich, eds. 1979. *A Source Book in Astronomy & Astrophysics*. Cambridge: Harvard University Press.

Laplace, Pierre-Simon. 1995. *Philosophical Essays on Probability* [1814]. New York: Springer Verlag.

Larson, Edward J., and Larry Witham. 1998. Leading Scientists Still Reject God. *Nature* 394: 313.

Lewis, John L. 1997. *Physics & Chemistry of the Solar System*. Burlington, MA: Academic Press.

Loomis, Elias. 1860. *An Introduction to Practical Astronomy*. New York: Harper & Brothers.

Lowell, Percival. 1895. *Mars*. Cambridge, MA: Riverside Press.

———. 1906. *Mars and Its Canals*. New York: Macmillan and Co.

———. 1909. *Mars as the Abode of Life*. New York: Macmillan and Co.

———. 1909. *The Evolution of Worlds*. New York: Macmillan and Co.

Lyapunov, A. M. 1892. *The General Problem of the Stability of Motion*. PhD thesis, University of Moscow.

Mandelbrot, Benoit. 1977. *Fractals: Form, Chance, and Dimension*. New York: W. H. Freeman & Co.

Maxwell, James Clerke. 1873. A *Treatise on Electricity and Magnetism*. Oxford, UK: Oxford University Press.

McKay, D. S., et al. 1996. Search for Past Life on Mars. *Science* 273, no. 5277.

Michelson, Albert A. 1894. Speech delivered at the dedication of the Ryerson Physics Lab, University of Chicago.

Michelson, Albert A., and Edward W. Morley. 1887. On the Relative Motion of Earth and the Luminiferous Aether. In *London Philosophical Magazine and Journal of Science* 24, 5th Series. Newcastle, UK: Literary and Philosophical Society.

Morrison, David. 1992. The Spaceguard Survey: Protecting the Earth from Cosmic Impacts. *Mercury* 21, no. 3: 103.

Nasr, Seyyed Hossein. 1976. *Islamic Science: An Illustrated Study*. Kent: World of Islam Festival Publishing Co.

Newcomb, Simon. 1888. *Sidereal Messenger* 7: 65.

———. 1903. *The Reminiscences of an Astronomer*. Boston: Houghton Mifflin Co.

[Newton, Isaac] David Brewster. 1855. *Memoirs of the Life, Writings, and Discoveries of Sir Isaac Newton*. London: T. Constable and Co.

Newton, Isaac. 1706. *Optice (Latin)*, 2nd ed. London: Sam Smith & Benjamin Walford.

———. 1726. *Principia Mathematica (Latin)*, 3rd ed. London: William & John Innys.

———. 1728. *Chronologies*. London: Pater-noster Row.

———. 1730. *Optiks*, 4th ed. London: Westend of St. Pauls.

———. 1733. *The Prophesies of Daniel*. London: Pater-noster Row.

———. 1958. *Papers and Letters on Natural Philosophy*. Ed. Bernard Cohen. Cambridge, MA: Harvard University Press.

———. 1962. *Principia Vol. II: The System of the World* [1687]. Berkeley: University of California Press.

———. 1992. *Principia Mathematica (English)* [1729]. Norwalk, CT: Easton Press.

Norris, Christopher. 1991. *Deconstruction: Theory & Practice*. New York: Routledge.

O'Neill, Gerard K. 1976. *The High Frontier: Human Colonies in Space*. New York: William Morrow & Co.

Planck, Max. 1931. *The Universe in the Light of Modern Physics*. London: Allen & Unwin Ltd.

————. 1950. *A Scientific Autobiography (English)*. London: Williams & Norgate, Ltd.

[Planck, Max] 1996. Quoted by Friedrich Katscher in The Endless Frontier. *Scientific American*, February, p. 10.

Ptolemy, Claudius. 1551. *Almagest* [c. 150]. Basilieae, Basel.

Salaam, Abdus. 1987. The Future of Science in Islamic Countries. Speech given at the Fifth Islamic Summit in Kuwait, http://www.alislam.org/library/salam-2.

Schwippell, J. 1992. Christian Doppler and the Royal Bohemian Society of Sciences. In *The Phenomenon of Doppler*. Prague.

Sciama, Dennis. 1971. *Modern Cosmology*. Cambridge, UK: Cambridge University Press.

Shamos, Morris H., ed. 1959. *Great Experiments in Physics*. New York: Dover.

Shapley, Harlow, and Heber D. Curtis. 1921. *The Scale of the Universe*. Washington, DC: National Academy of Sciences.

Sullivan, W. T. III, and B. J. Cohen, eds. 1999. *Preserving the Astronomical Sky*. San Francisco: Astronomical Society of the Pacific.

Taylor, Jane. 1925. *Prose and Poetry*. London: H. Milford.

Tipler, Frank J. 1997. *The Physics of Immortality*. New York: Anchor.

Tucson City Council. 1994. *Tucson/Pima County Outdoor Lighting Code*, Ordinance No. 8210. Tucson, AZ: International Dark Sky Association.

[Twain, Mark] Kipling, Rudyard. 1899. An Interview with Mark Twain. *From Sea to Sea*. New York: Doubleday & McClure Company.

Twain, Mark. 1935. *Mark Twain's Notebook*.

van Helden, Albert, trans. 1989. *Sidereus Nuncius*. Chicago: University of Chicago Press.

Venturi, C. G., ed. 1818. *Memoire e Lettere*, vol. 1. Modena: G. Vincenzi.

von Braun, Werner. 1971. *Space Frontier* [1963]. New York: Holt, Rinehart and Winston.

Wells, David A., ed. 1852. *Annual of Scientific Discovery*. Boston: Gould and Lincoln.

White, Andrew Dickerson. 1993. *A History of the Warfare of Science with Theology in Christendom* [1896]. Buffalo, NY: Prometheus Books.

Wilford, J. N. 1999. Rarely Bested Astronomers Are Stumped by a Tiny Light. *The New York Times*, August 17.

Wright, Thomas. 1750. *An Original Theory of the Universe*. London: H. Chapelle.

NAME INDEX

SUBJECT INDEX